A COMPREHENSIVE GUIDE TO

SIGN LANGUAGE INTERPRETING IN EUROPE

2026 EDITION

Maya de Wit, PhD

ISBN 9789403895536

First published 2026

Publisher and author: Maya de Wit

Front cover: Vivian van Schagen

Suggested citation:

Wit, M. de (2026). A Comprehensive Guide to Sign Language Interpreting in Europe, 2026 edition. Self-published. Printed by Bookmundo, Baarn, M. de Wit

Contents

Foreword 5
Summary 7
List of respondents by country or region 8
Introduction 9
Part 1 - Associations 11
1.1 Establishment of sign language interpreter associations 12
1.2 National associations 13
1.3 Names of the associations 15
1.4 Countries without an independent national interpreter organization 16
1.5 Membership ratios 16
1.6 Membership numbers over the years 18
1.7 Representation 21
1.8 Deaf interpreters as a member 23
1.9 Membership requirements 25
1.10 Membership benefits 28
1.11 Aims of the association 30
1.12 Collaboration with the deaf organization 31
1.13 Funding of the association 31
1.14 Membership fees 31
1.15 Associations' volunteer and paid positions 34
1.15 Other types of membership 34
1.16 Main issues the associations are facing 35
Part 2 - Quality assurance 39
2.1 Quality assurance 40
2.2 Definition in national legislation 42
2.3 Regulation of the sign language interpreter profession 45
2.4 Registry of interpreters 46
2.4 Complaints' procedure 52
Part 3 - Deaf Interpreters 59
3.1 Deaf interpreters - terminology 60
3.2 Deaf interpreter developments 60
3.3 Hearing interpreters supporting deaf interpreter developments 64
3.4 Deaf interpreters in official standards 67
3.5 Measures to strengthen deaf interpreters' position 69
3.6 Work and employment of deaf interpreters 72
3.7 Training and education of deaf interpreters 81
3.8 Current main barriers and future steps 88

Part 4 - Training & Education 89
4.1 Educational Programs 90
4.2 Countries and regions without a program 92
4.3 Past and new programs 94
4.4 Prerequisites to attend the program 107
4.5 Interpreting for deafblind persons 112
4.6 Interpreter training in specific topics 113
4.7 Interpreting degree as a prerequisite to work 115
4.8 Continuing education 116
4.9 Interpreter specializations 118
4.10 Mentoring 120
Part 5 - Employment 123
5.1 Interpreters currently working 124
5.2 Part-time versus full-time 128
5.3 Interpreting for deafblind persons 129
5.4 Lack of interpreters 131
5.5 Number of working interpreters: 2001 - 2025 132
5.6 Number of deaf signers 133
5.7 Deaf signers per interpreter 134
5.8 The right to an interpreter 138
5.9 Requesting interpreting services 144
5.10 Party responsible for the payment of interpreting services 147
5.11 Recognition of sign language 153
5.12 Forms of employment 154
5.13 Onsite and remote interpreting 156
5.14 Freelance interpreting fees 157
5.15 Payment for specific settings or competencies 162
5.16 Interpreting for deafblind persons 164
5.17 Reasonable pay 168
5.18 Interpreting as a main occupation 170
5.19 Budget limitations 171
5.20 Working in paid employment 172
5.21 Development of the profession 172
5.22 Working conditions - development 172
5.23 Impact of the Covid19 pandemic on the profession 175
5.24 Impact of AI on the profession 177
5.25 Main professional opportunities and challenges in 2025 179
Appendices 185
Appendix 1 - Survey sign language interpreting in Europe 186
Appendix 2 - Overview of facts by country 199
Appendix 3 - Overview of all tables and figures 213

Foreword

I am delighted to share with you the outcomes of the seventh European survey, six years after the last edition. As before, I have conducted this as a pro bono project, depending on colleagues across Europe who, like myself, donated their time to advance the profession as a whole. I could not have done this without you. Thank you for contributing to this collaborative achievement.

I especially would like to thank the European Union of the Deaf (EUD) for the reflective exchanges and collaboration and the European Forum of Sign Language Interpreters (efsli) for their input. Thank you also to all the individual interpreters who provided suggestions for additional questions to the survey. A special thank you goes out to my colleagues Sandra Schügerl for her professional review of the survey questions, and Tina Vrbanic for her thoughtful support in the analysis. Sadly, our colleague and friend Anna Komorova, who was my lifeline to some of the Eastern European countries for the last editions, passed away in October 2022. She is sorely missed. I am incredibly thankful for Victoria Savostenkova, who stepped in and translated the survey into Russian, reached out to our colleagues in Eastern Europe, and translated all the answers back to English.

This edition required an immense investment in time and effort. The survey platform I had used previously deleted my European surveys and all their data, unannounced. I had to find a new platform and rebuild the survey with over 300 questions and the corresponding logic. The patience of the respondents and myself was tested by the new survey platform, which turned out to be less robust than promised. The analysis of the data came also with major setbacks as the new platform was not up to the complexity of the survey. The time needed to fix the errors and finalize the analysis forced me to publish this book later than planned.

That said, I am much pleased to present you now with the latest, most comprehensive data collection on the status of the sign language interpreting profession in Europe.

Maya de Wit, PhD

Summary

The profession of sign language interpreting has changed significantly since the previous edition of this book was published in 2020. The Covid19 pandemic has resulted in greater awareness among the public due to the increased visibility of sign language interpreters in the media. The pandemic also furthered the use of technology and remote interpreting. In addition, the number, duration, and levels of educational programs for sign language interpreters have increased. New university programs have been established in Slovakia and Poland, and the deaf interpreter program in Sweden has now become permanent. Overall, there are more training and work opportunities for deaf interpreters, which, however, are not yet on par with those for hearing interpreters.

The number of professionally trained sign language interpreters continues to grow, with nearly 11,500 interpreters across the forty-five countries and regions covered in this survey. There are improvements in the working conditions, with adequate agreements and better pay. Still, many countries see a shortage of interpreters, for instance, in remote regions or when interpreters with special skills, such as additional languages, are required.

The lack of professional recognition is one of the major issues that hinder further development of the profession and is often linked to funding. Consequently, deaf signers are faced with interpreting services that are insufficient in terms of quality, quantity, or both. Overall, the profession needs sustainable national infrastructures to match the growing demand for qualified sign language interpreters.

List of respondents by country or region (N=47)

Albania[3]

Austria[1]

Belgium-Flanders[1]

Belgium-Wallonia[1]

Bosnia-Herzegovina[HI]

Bulgaria[1]

Croatia[1]

Cyprus[1]

Czech Republic[1]

Denmark[1, DI]

England, Wales & Northern Ireland (ASLI & NUBSLI)[2]

Estonia[1]

Finland[1]

France[1]

Georgia[3]

Germany[1,DI]

Greece[1]

Hungary[1]

Iceland[1]

Ireland[1]

Italy[2] (ANIOS & ANIMU)

Kosovo[1]

Latvia[5]

Lithuania[1]

Luxembourg[HI]

Malta[1]

Moldova[3]

Montenegro[3]

Netherlands[1]

Norway[1]

Poland[1]

Portugal[1]

Romania[4]

Russia[1]

Scotland[1]

Serbia[1]

Slovakia[1]

Slovenia[1]

Spain[1]

Sweden[1]

Switzerland-French region[1]

Switzerland-German region[1]

Switzerland-Italian region[1]

Turkey[HI]

Ukraine[HI,3]

[1]*National interpreter association*
[2]*Two national interpreter associations*
[3]*National deaf association*
[5]*Interpreting services center*
[HI]*Individual response (hearing interpreter)*
[DI]*Individual response (deaf interpreter)*

Introduction

The seventh edition of *Sign Language Interpreting in Europe* provides an overview of the current situation of the sign language interpreting profession. The data in this publication is based on a European survey which was sent out and collected in the autumn of 2025.

A total of forty-seven respondents (two associations from Italy and two from England, Wales and Northern Ireland) covering forty-five regions and countries, responded to the survey. This publication presents responses from nearly all forty-seven members of the Council of Europe (CoE).[1] For the first time, Moldova participated in this European survey.

The European Union of the Deaf (EUD), the European Deafblind Union (EDbU), and the European Union of the Deaf Youth (EUDY), European Forum of Sign Language Interpreters (efsli), as well as interpreter associations and individual stakeholders, were asked to provide input for the survey. Back in 2001, when I conducted this survey for the first time, nineteen countries participated. The data from the subsequent surveys (2001, 2004, 2007, 2012, 2016, 2020 and now 2025) show which trends shaped the sign language interpreting profession during these years.

The survey consists of questions concerning the interests of interpreters in areas such as education, employment, remuneration, and legislation. The survey was sent to all national associations of sign language interpreters in the respective countries. In countries which do not have an association for sign language interpreters, or where there was no response from the interpreter association, the survey was answered by the national deaf association or an individual expert in the field. A list of all the respondents can be found on page 8.

For the first time ever, this edition provides a chapter on the development of the profession for deaf interpreters. In addition, the publication focuses on the functioning of the professional associations, and how current developments can impact the profession. New information is also provided on continuing education, specializations, and mentoring of sign language interpreters.

This publication consists of five parts: associations, quality control, deaf interpreters, education, and employment. The appendices provide the full survey and 'Facts and figures per country'. Additional links and information are published on https://www.mayadewit.nl/european-study.

The outcome of this survey depends on people volunteering their time to answer many questions, some of which are hard to estimate. Where possible, the survey allows respondents to skip answers by choosing "I do not know." Still, human error in a survey of this scale is unavoidable and at times a data point may be off the mark. That said, the sheer scale of this survey, covering the entire geography over a time span of twenty-five years, should be helpful in identifying leading and emerging trends in the field.

[1] https://www.coe.int/en/web/portal/members-states (accessed December 2025)

Part 1 – Associations

1.1 Establishment of sign language interpreter associations

National sign language interpreter associations in Europe represent a country or a language region. Most of the respondents indicate that they have a national association for sign language interpreters in their country or region. For an overview of all the associations, please refer to Appendix 1, 'Facts and figures per country.'

Table 1.1 lists the associations that are officially independent interpreter associations and when they were established. Listed are those associations that responded or whose data is known. Some countries or regions are listed more than once, because associations changed their names or folded and a new one was established.

The first national association of sign language interpreters was established in Sweden in 1969. Denmark and Norway followed in 1977 and 1978, respectively. In 1978 France was the first non-Scandinavian country to set up an organization. Estonia was officially registered in 1995 but established in 1994[2]. Since the last publication in 2020 no new national associations have been established.

In Germany there are several regional associations and two national associations, BGSD and tgsd. The BGSD has 11 regional associations.[3] Tgsd, the national organization of deaf interpreters in Germany established by and for deaf interpreters, did not participate in the survey this time, but an individual qualified deaf Interpreter (DI) who is a member of tgsd did. Other countries with an independent interpreter association and regional associations are Spain and Russia.

For the first time NUBSLI (National Union of British Sign Language Interpreters), another national association in England Wales & Northern Ireland,[4] participated in the survey. Romania no longer has a national association of sign language interpreters. An individual representative provided input instead.

[2] Liivi Liholm, personal communication (7 December 2020)
[3] https://bgsd.de/de/verband/arbeitsgruppen.html
[4] For England, Wales & Northern Ireland the abbreviation 'England, W&NI' will be used in this publication.

Table 1.1 Year of establishment of independent interpreter associations					
Country	**Name**	**Established**	**Country**	**Name**	**Established**
Sweden	STTF	1969	**Czech Republic**	CKTZJ	2000
Denmark	FTT	1977	**Spain**	FILSE	2000
Norway	TiA[5]	1978	**Belgium-Flanders**	VVTG	2001 - 2009
France	AFTILS[6]	1978	**Slovenia**	ZTSZJ	2002
Scotland	SASLI	1981 - 2019	**Bulgaria**	НАПЖЕБ	2005
Finland	SVT	1982 - 2019	**Slovakia**	SKTPJ	2007
Belgium-Flanders	BVDT	1983 - 2001	**Croatia**	HDTPZJG	2008
Belgium-Wallonia	ABILS	1986	**Kosovo**	KASLI	2008
Switzerland-Fr.	ARILS[7]	1986	**Belgium-Flanders**	BVGT	2009
Ireland	IASLI	1987 - 2007	**Russia**	RASLI	2009
England, Wales & NI	ASLI	1987	**Lithuania**	LGKVA	2009
Italy	ANIOS	1987	**Poland**	STPJM	2009
Italy	ANIMU	1987	**Serbia**	ATSZJ	2010
Netherlands	NBTG[8]	1988	**Ireland**	CISLI	2011
Greece	SDENG	1991	**Portugal**	ANAPI-LG	2011
Switzerland-Ger.	BGDÜ[9]	1991	**Germany**	tgsd	2011
Hungary	JOSZ	1994	**England, Wales & NI**	NUBSLI	2014
Estonia	EVK	1994	**Cyprus**	ΣΔΚΝΓ	2015
Germany	BGSD	1997	**Croatia**	CASLIFD	2016
Austria	ÖGSDV	1998	**Finland**	Kieliasiantun tijat	2019
Switzerland-It.	ILISSI	1998	**Scotland**	SCOSLI	2020
Iceland	HART	1998	**Malta**	AILSM	2020

1.2 National associations

Some countries have more than one national association, such as Belgium, Portugal, Switzerland, Ukraine, and the United Kingdom. Not having a single national association per country is typically caused by difference in language, culture, geography, and regional regulations. For instance, if interpreters do not share the same (spoken) language, communicating and working together in one organization can be challenging.

Switzerland officially has four spoken languages: German, French, Italian, and Rhaeto-Romance. The three associations for sign language interpreters (ARILS, BGDÜ & ILISSI) in

[5] Name change in 2020.
[6] In January 2024, the association changed its name and became AFTILS: French association of sign language translators and interpreters (source: https://aftils.fr/histoire-de-l-aftils).
[7] The association 'ARILS' was founded in 1986 under the name AILS (source: http://arils.ch/historique/).
[8] In 2000 the name changed from NVTD to NBTG.
[9] Name change in 2024, source: https://www.bgdu.ch/geschichte

Switzerland are linked to the regions of Switzerland where German Sign Language (BGDÜ), French Sign Language (ARILS), and Italian Sign Language (ILISSI) are used.

Belgium also has more than one official spoken language, namely French, Dutch, and German. Since there is such a linguistic difference between Flanders and Wallonia, anything pertaining to sign language interpreting is taken care of separately.[10] The sign language interpreters in the Flemish part of Belgium have an organization called BVGT. In Wallonia the interpreter association is named ABILS.

Most national associations for sign language interpreters in Europe are directly linked to a single national sign language rather than to a single European country. If a country has more than one spoken language, it has at least an equivalent number of national sign languages. The exceptions are Scotland and England, Wales & Northern Ireland. They all use British Sign Language, yet Scotland has its own distinct regional organization. Therefore, when referring to the United Kingdom this publication distinguishes Scotland from England, Wales & Northern Ireland.

A country that has a single national sign language and more than one national association, is Portugal. Portugal has two associations which is reported as positive by ANAPI-LG, one of the national associations:

> *"Two associations with different objectives and are located in different parts of the country. While one is more focused on providing interpretation services and operates mainly in the north of the country, the second emerged with the purpose of defending professionals in the field and investing in training. Currently, members are divided between the two associations depending on their goals."*

A similar answer is provided by the respondent from Turkey:

> *"Having more than one association can be beneficial for deaf individuals and interpreters, as it allows better response to diverse needs and regional demands. However, without coordination and shared goals, it may lead to fragmentation and inefficiency."*

Italy is the country with the most national associations, namely seven. This is due to several national laws that passed in 2021[11] and in subsequent years the official recognition of Italian Sign Language (LIS) and the profession of signed language interpreting. Associations of sign language interpreters in Italy are responsible for validating skills and issuing qualification certificates.[12] As a consequence interpreters established associations to issue these validations.

[10] M. van Herreweghe & M. van Nuffel (2000). RID Journal of Interpretation: Sign (Language) Interpreting in Flanders, Belgium
[11] https://www.gazzettaufficiale.it/eli/id/2021/05/21/21A03181/sg
[12] F. Malaspina (2025). Moving Towards Professional Standards: An Investigation into the Role and Quality of the Signed Language Interpreting Profession in Italy. Master Thesis, University of Siena, Italy.

Croatia also has three associations, which is not perceived as advantageous by CASLI:

> *"One of the three associations operates exclusively in a single city, while another does not include the term 'sign language interpreter' in its name, instead using 'communication intermediary.'*
>
> *The Croatian Association of Sign Language Interpreters was established by interpreters working within various deaf associations and does not employ interpreters on a full-time basis. Until last year, the Association was unable to employ any SLIs due to the absence of an office. Moreover, to operate under the Ministry of Labor, Pension System, Family and Social Policy, it is necessary to meet demanding and costly requirements."*

Until 2019 sign language interpreters in Finland had an association (SVT) of sign language interpreters only. As of 2019 Finland has a new national association called Kieliasiantuntijat ry (language experts) that includes interpreters, translators and other professionals working in the field of multilingual communications. This includes spoken languages as well as signed languages.

Scotland established their first association (SASLI) in 1981. However, in 2018 the government decided to withdraw funding for SASLI. As SASLI practiced a dual function of register and membership organization, the Scottish government felt that this was a conflict of interest. Through a vote by the SASLI members it was decided that SASLI would become a registering-only body. This new organization was named the SRLPDC (Scottish Register of Language Professionals with the Deaf Community) and was officially confirmed in 2020. It is also referred to as 'The Scottish Register.' In addition, a separate membership organization for sign language interpreters SCOSLI (Scottish Collaborative of Sign Language Interpreters) was established in 2020.

1.3 Names of the associations

Most of the independent associations of sign language interpreters are called organization or association of sign language interpreters and have added their nationality to it. For instance, the Dutch association is named 'The Dutch Association of Sign Language Interpreters.'[13] In France the association changed its name from AFILS to AFTILS to include translators in the name. In Finland the association is not exclusively for sign language interpreters, and the umbrella term 'language experts' is used in the name.

In the past the Spanish association FILSE added 'Guide-Interpreters' to their name: 'Spanish Federation of Spanish Sign Language Interpreters and Guide-Interpreters.' AFTILS (France) and ÖGSDV (Austria) refer to both interpreters and translators in their association names. All the other associations refer to their national sign language in the name of the association.

[13] Nederlandse Beroepsvereniging Tolken Gebarentaal (NBTG).

1.4 Countries without an independent national interpreter organization

The countries which do not have an independent national association of sign language interpreters are Albania, Bosnia and Herzegovina, Georgia, Latvia, Luxembourg, Moldova, Montenegro, Romania and Ukraine. Some of these countries, such as Latvia and Ukraine, have a sub-organization or committee for interpreters within or under the national deaf organization. In Ukraine that is the Council of Sign Language Interpreters, an advisory and consultative structural unit of the Ukrainian Society of the Deaf, which does not provide for membership or separate regional branches but is like a labor union for interpreters working for the national association of the deaf. Recently, a new organization was established: the All-Ukrainian Association of Sign Language Interpreters and People with Disabilities. This association does not have a national status nor formal membership criteria and has fewer than fifty members in Ukraine.

North Macedonia did not participate in the survey but some information on this country was available online. According to the national deaf association[14] there are thirty-eight licensed interpreters and there is no national interpreter association.[15]

1.5 Membership ratios

One of the questions in the survey concerns the percentage of sign language interpreters in the country or region which are a member of their national association. In total 37 respondents were able to answer this question on behalf of their (national) associations. Romania was not able to answer this question as the total number of sign language interpreters in the country is unknown. In 2016 Romania estimated membership at five percent.

In the surveys in 2007, 2012, and 2016 the three regions in Switzerland had the highest percentage. Most of the interpreters were a member of their affiliated association. This changed in 2020 as Slovenia, Malta and Switzerland-French region now top the list with membership ratios of hundred percent. Greece and Kosovo also moved towards the top of the list.

Figure 1.1 indicates the percentage of interpreter members per country or region. The countries that do not have an organization of sign language interpreters or did not have the data are not listed in this figure. Several countries answered with an estimation. If the given estimation was a range, the average was taken to provide an indication of the actual numbers.

---

[14] https://deafmkd.org.mk/en/list-of-interpreters/

[15] Teuta Rasimi, assistant professor at the State University of Tetovo, North Macedonia, personal communication (28 November 2025)

Fig. 1.1 Sign language interpreters and national organizations: relative membership (%)

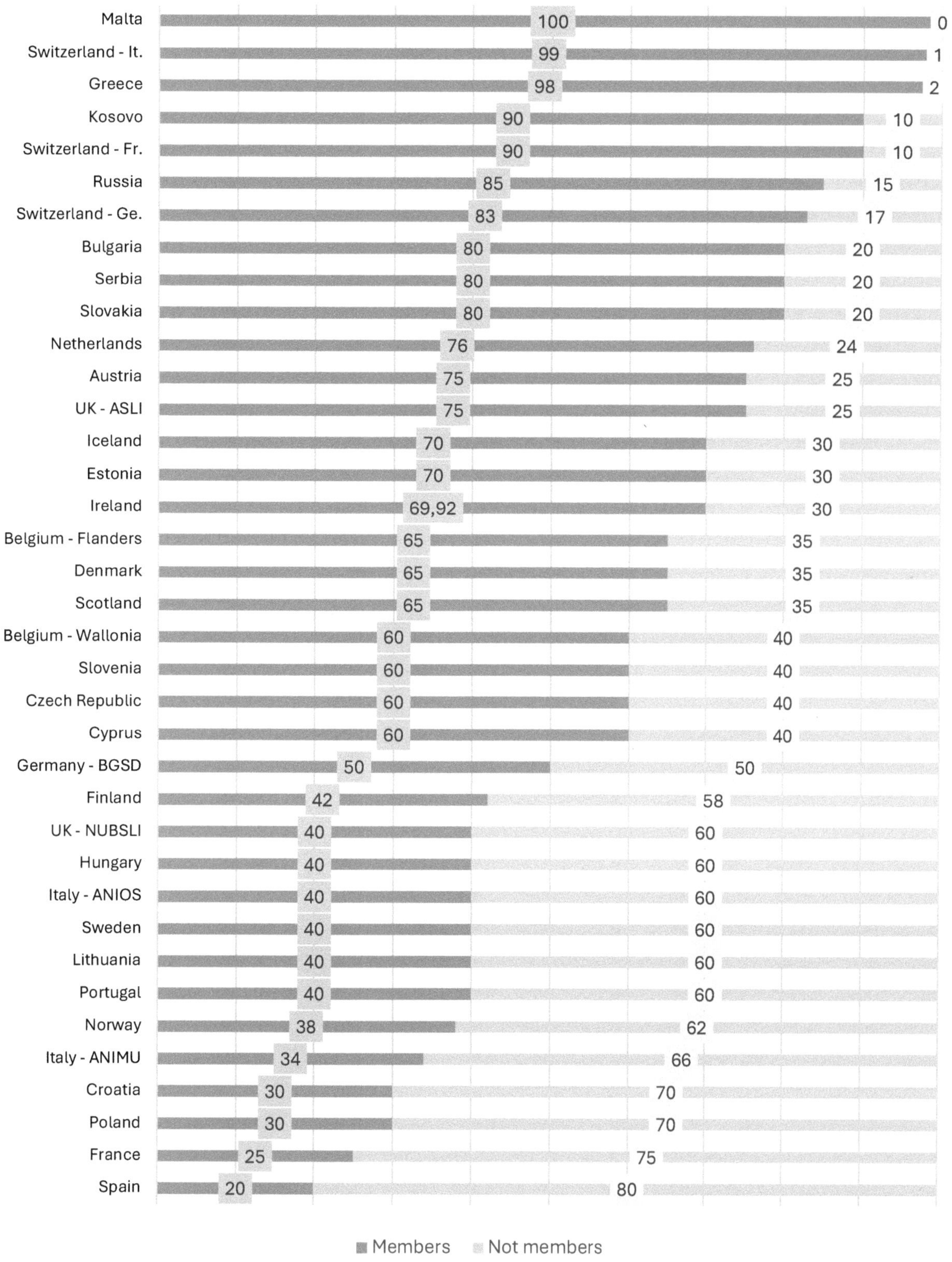

1.6 Membership numbers over the years

A total of thirty-seven respondents filled out the survey on behalf of their sign language interpreter associations. Figure 1.3 shows the number of members per association in 2016, 2020 and 2025. ASLI in England, Wales & NI has the highest number with 1340 interpreter members and reports the highest increase compared to 2016 and 2020.

The associations in the smaller countries and regions have the least number of interpreter members, such as Cyprus (22), Estonia (17), Malta (13), and Iceland (20). The biggest decrease in membership was for Finland.

Fig. 1.2a Total membership numbers by national or regional association (100 members >)

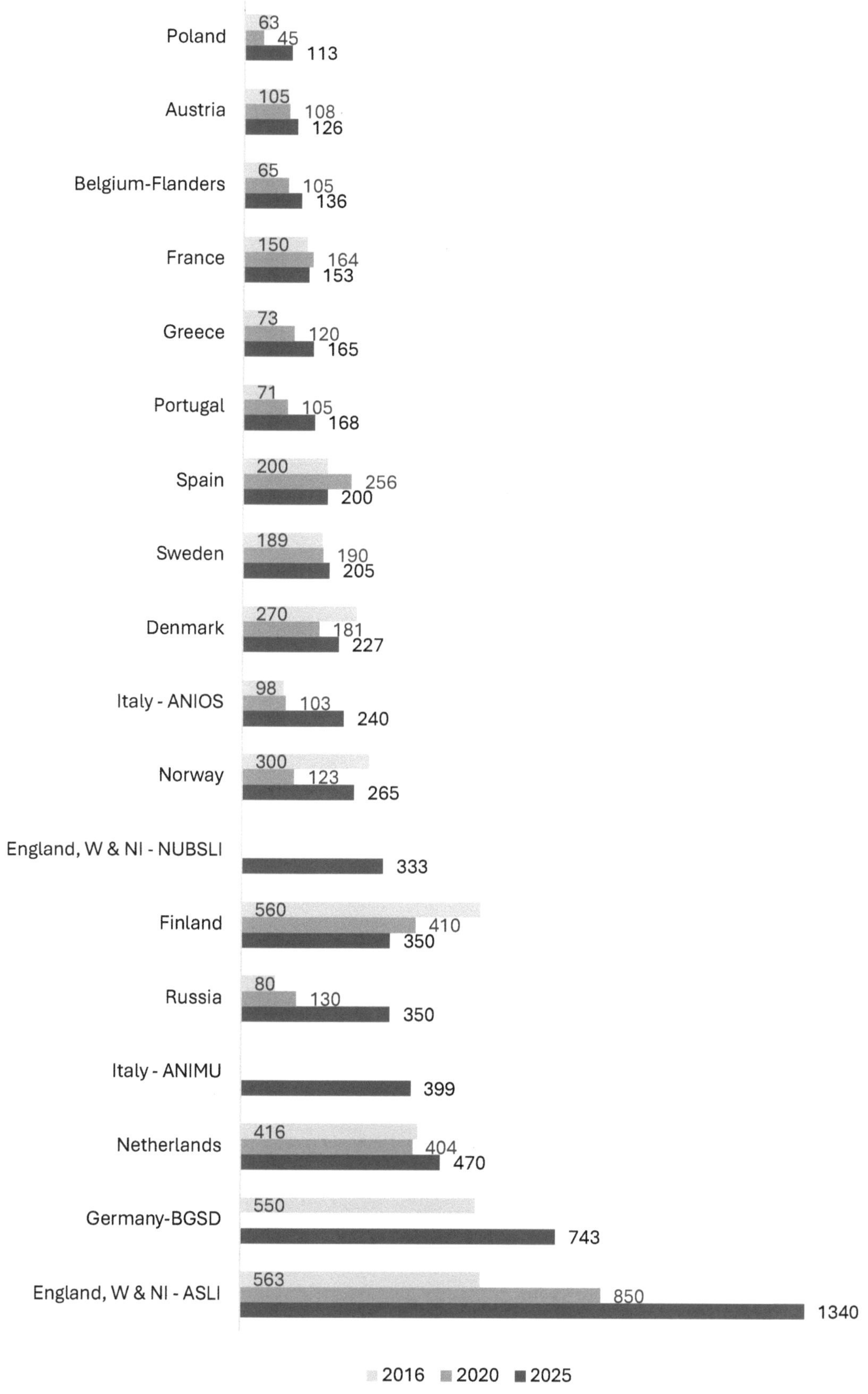

Fig. 1.2b Total membership numbers by national or regional association (<100 members)

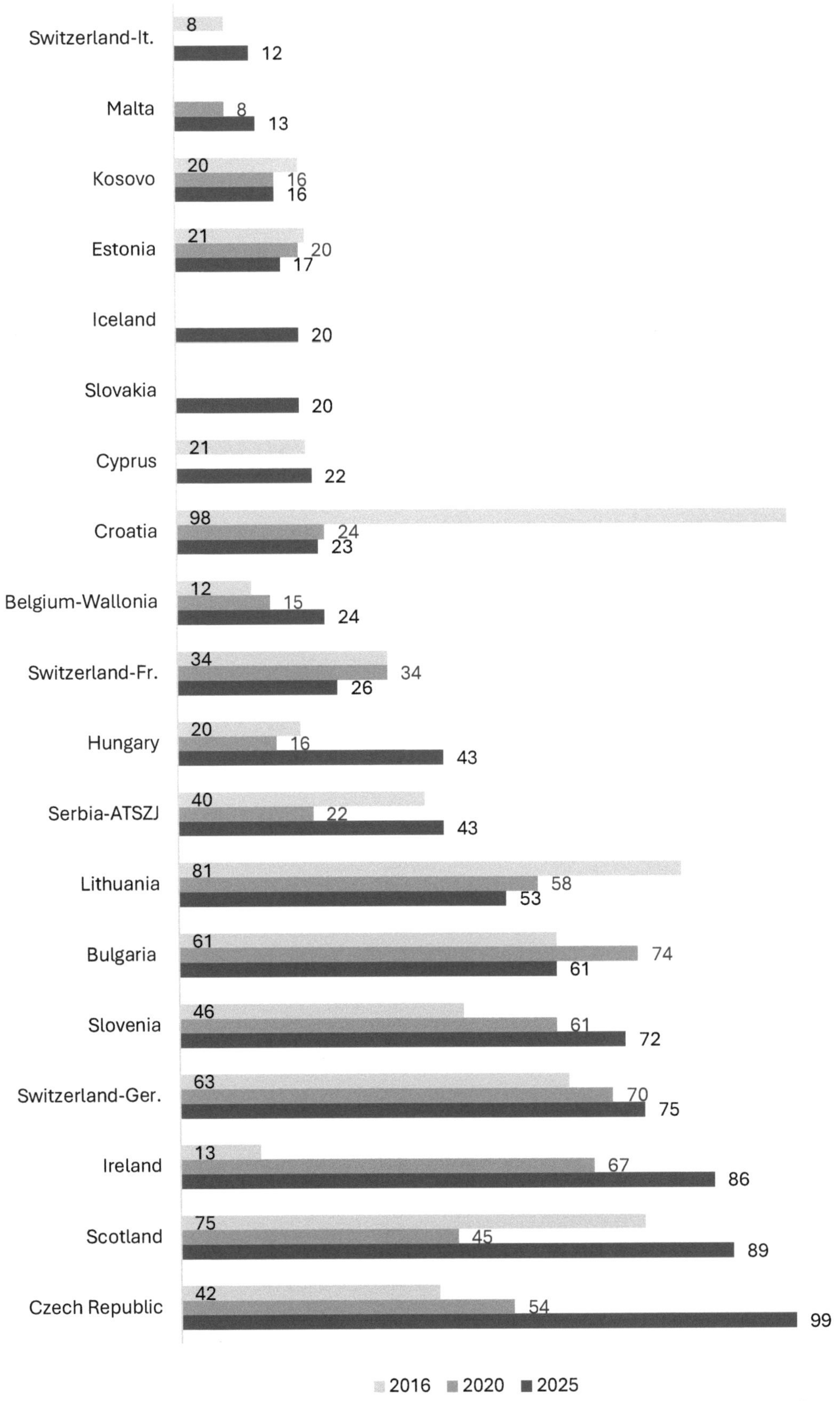

1.7 Representation

Out of the thirty-seven national interpreter associations, thirty gave an estimate on the percentage of gender distribution in the association (Fig. 1.3). The percentage of female members has remained approximately the same during the last 10 years and is on average almost ninety percent. The percentage of female members in an association is reported to be at least eighty percent in all the associations.

Fig. 1.3 Percentage of gender distribution among the members of all interpreter associations (n=30)

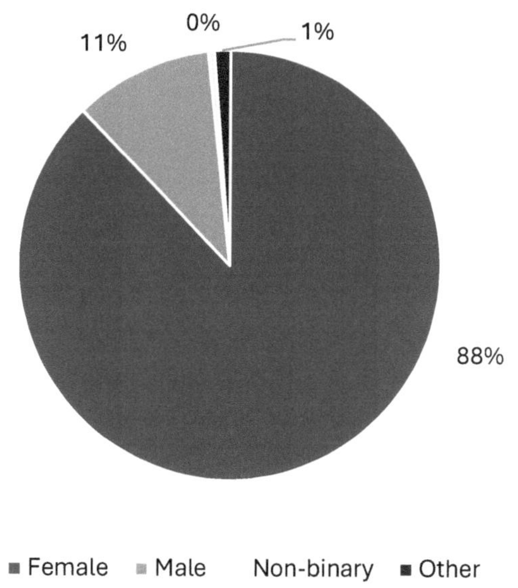

In this survey for the first time the question was asked if the interpreter association collects data on gender and representation of ethnic and racial minorities in their association by asking members to self-identify. Figure 1.4 shows that most of the associations do not collect this information. When asked if they belief that racial and ethnic minorities are represented in their interpreter association, the majority (62%) indicates that this is not the case (Fig. 1.5).

Fig. 1.4 Does your association collect member information on gender, race and ethnicity by asking members to self-identify? (n=37)

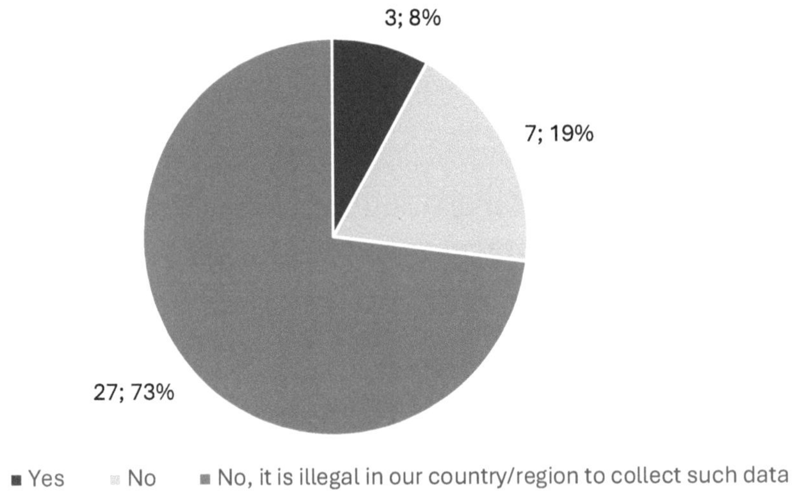

Fig. 1.5 Representation of ethical and racial minorities in the interpreter associations (n=37)

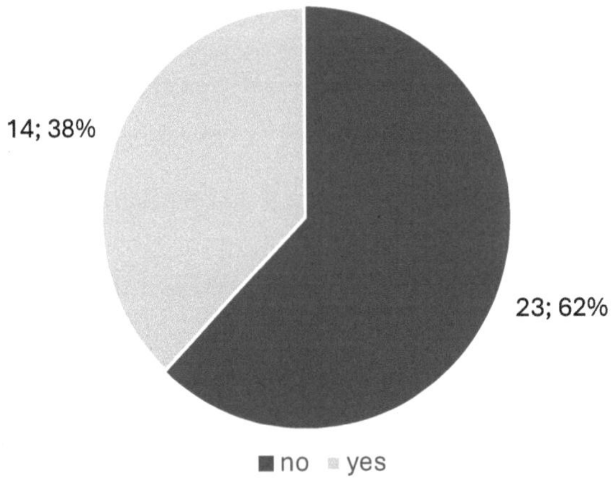

AFTILS (France) reports that in France, institutional statistics about ethnic and racial minorities are not collected, because of the mainstream perspective of universalism. Therefore, AFTILS does not have any data in in their association, nor specific board or working group for representation.

The BGSD (Germany) state that all are members:

> *"It is a topic we need to consider more strongly. We would like to represent all working SLI. What will make them feel represented?"*

ASLI (England, W&NI) mentions that they have members from various ethnic minorities and characteristics, but do not collect this data. However, ASLI did commission a census in the United Kingdom on sign language interpreters and translators, including questions on nationality and heritage, which was published in 2022[16].

1.8 Deaf interpreters as a member

Deaf interpreters[17] work in a variety of settings, interpreting in a team with hearing sign language interpreters, interpreting between two signed languages, such as a national sign language and international sign,[18] and translating between written text and sign language. Examples of potential interpreting situations, provided by the respondents, involve consumers with minimal language skills, legal settings, deaf persons who use a foreign sign language, interpreting for deafblind persons (e.g. tactile), and mental health settings.

In 2020 a total of eighteen associations provided numbers on how many members are deaf. In 2025 there are twenty-seven national associations who report to have interpreter members who are deaf. Figure 1.6 indicates the membership of deaf and hard of hearing interpreters in those countries.

The countries and regions who report to have no deaf interpreters as members are Cyprus, Finland, Greece, Hungary, Iceland, Kosovo, Malta, Norway, Switzerland-French region, and Sweden. Finland mentions that they do not collect information about members' hearing status and therefore cannot report on the number of deaf and hearing members.

[16] Napier, J., Skinner, R., Adam, R., Stone, C., Pratt, S., Hinton, D. P., & Obasi, C. (2022). Representation and diversity in the sign language translation and interpreting profession in the United Kingdom. *Interpreting and Society: An Interdisciplinary Journal*, 2(2), 119-140. https://doi.org/10.1177/27523810221127596
[17] http://gupress.gallaudet.edu/excerpts/DIAW.pdf
[18] Wit, M. de (2023). Practitioners' perspectives on International Sign Conference Interpreting.

Fig. 1.6 Number of deaf and hard of hearing interpreters as members of an interpreter association

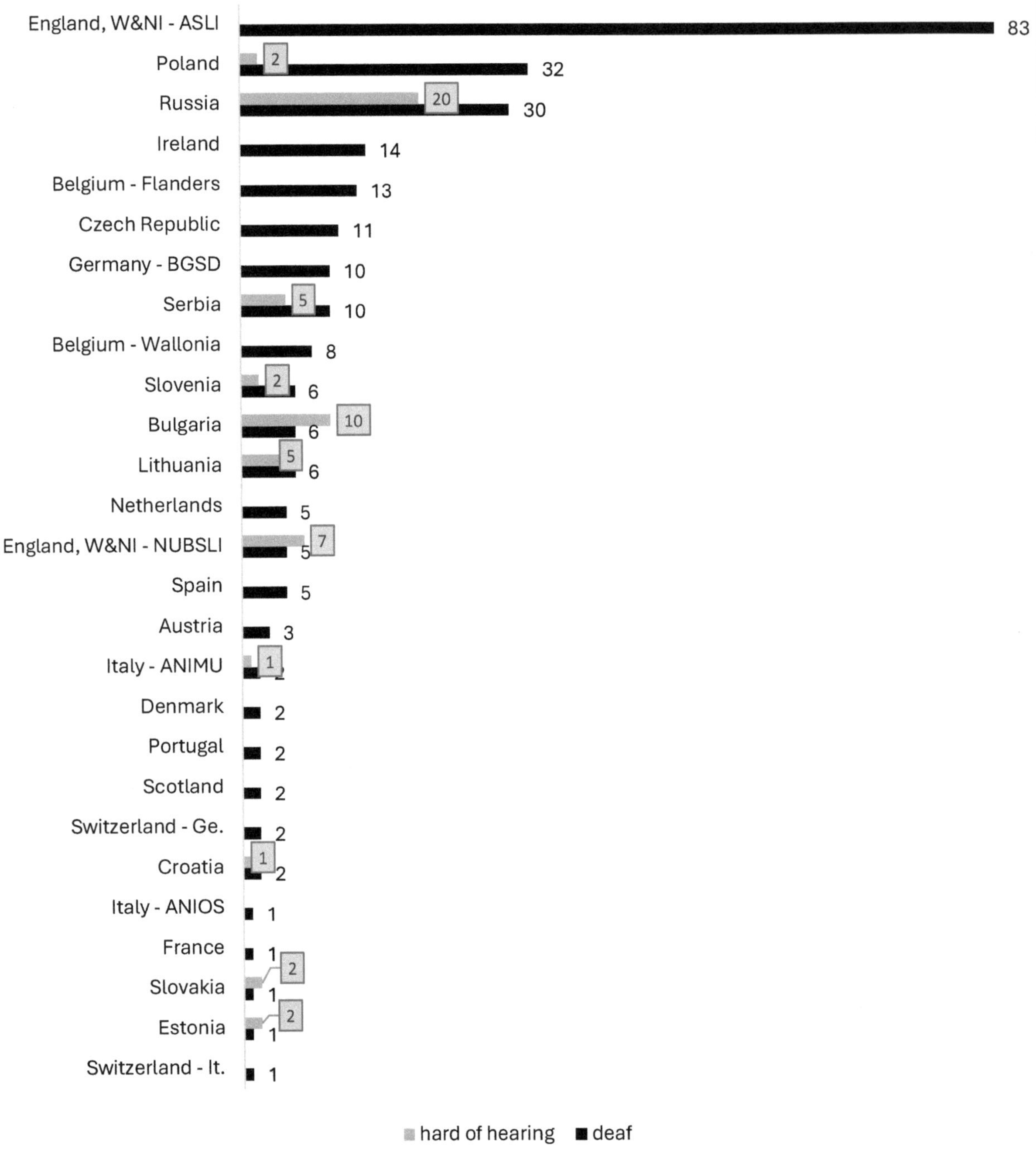

1.9 Membership requirements

To become a member of a national or regional sign language interpreter association, candidates must fulfill certain criteria. Most associations indicate that a diploma or a degree of an interpreting program are the most important membership requirements (Fig. 1.7). Some associations have different membership types, for example for students or trainers. This is detailed in paragraph 1.15.

Fig. 1.7 Individual membership requirements

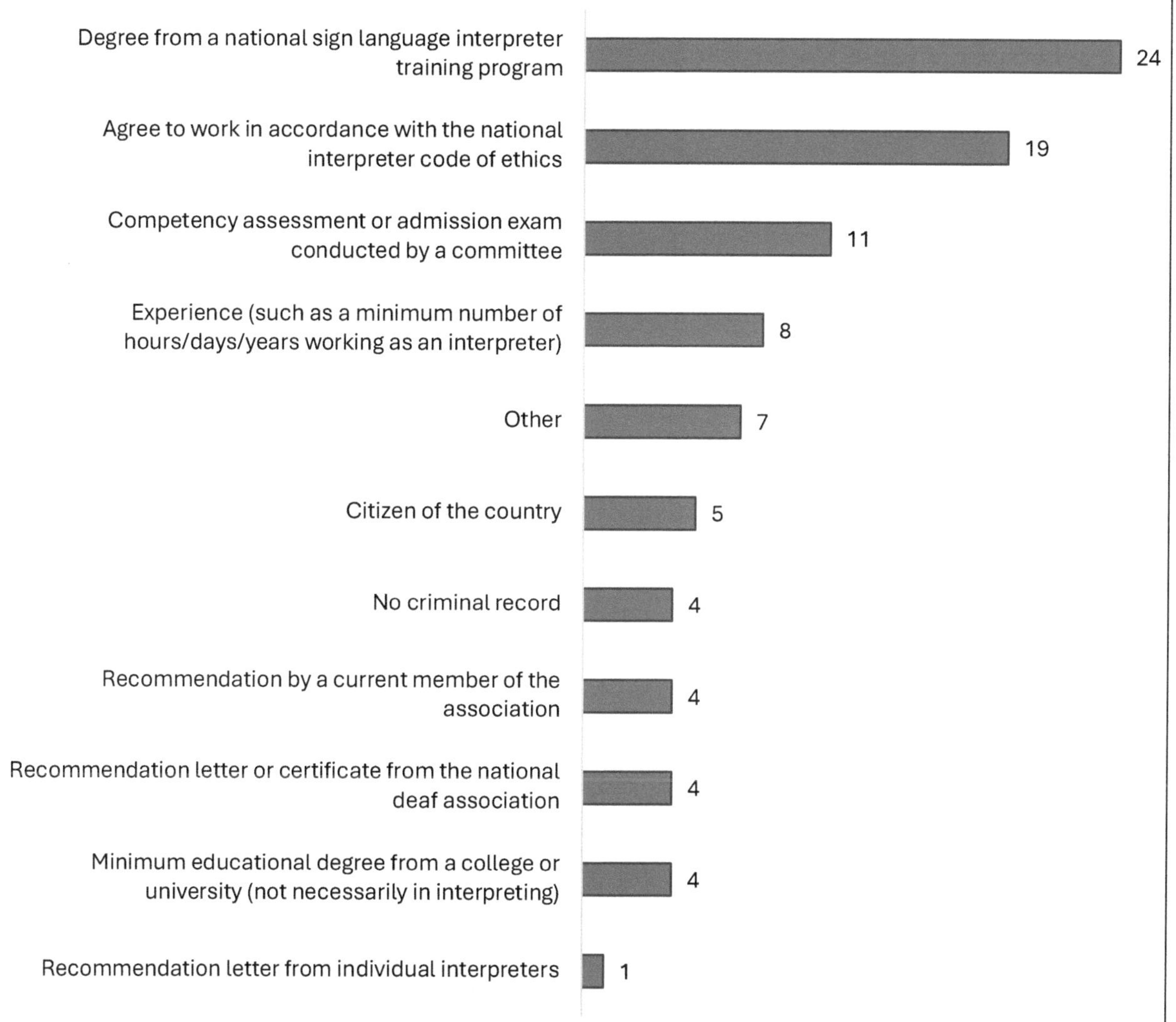

The landscape of sign language interpreter certification and membership requirements varies significantly across European countries, reflecting different stages of professionalization and regulatory development within the field. Under 'other' several associations detailed more requirements:

Slovakia - SKTJP: The Slovak Association of Sign Language Interpreters (SKTPJ) is currently undergoing a significant period of transformation following a period of inactivity from 2020 to 2024 due to the COVID19 pandemic. The organization is being revitalized with substantial changes to its statutes, code of ethics, and membership requirements. Historically, Slovakia's interpreting landscape has been characterized by a lack of formal education and regulation:

> "Until 2020, approximately 30 interpreters practiced without formal qualifications, as no university-level training programs existed. The launch of the first BA interpreting program for both hearing and deaf interpreters at Trnava University in 2020 marked a watershed moment for the profession. Prior to this development, the field was dominated by Children of Deaf Adults (CODAs) who provided interpreting services in a largely unregulated manner. A second association, ATLPRENES, issued certificates until approximately 2012-2013, though these required only "knowledge" of sign language without rigorous assessment. Consequently, SKTPJ maintained an open membership policy without formal requirements, though new standards are currently being developed pending board approval and ratification at the General Assembly."

Ireland (CISLI): Ireland operates through the Register of Sign Language Interpreters of Ireland (RISLI), which maintains a publicly accessible register of qualified interpreters. Membership requires recognized qualifications as specified by RISLI's standards, ensuring that all practicing interpreters meet established professional criteria.

Finland - Kieliasiantuntijat: The Finnish association requires members to either work as sign language interpreters, hold a degree in interpretation, or be actively studying interpretation, ensuring that membership is limited to those with formal training or active engagement in professional development.

England, Wales & Northern Ireland - ASLI: In the United Kingdom all members are required to be registered with appropriate regulatory bodies, which conduct comprehensive checks to ensure professional standards are maintained.

Slovenia: Established a rigorous national certification system administered through a state examination center. Prospective interpreters must pass a national professional qualification, preceded by an extensive portfolio assessment process. A consultant evaluates candidates' suitability and collects evidence, recommendations, and assessments before they appear before the examination commission. The Slovenian system requires knowledge extending beyond sign language proficiency to encompass understanding of deaf and hard of hearing community organization and functioning, as well as national legislation in social welfare. Interpreters also have the option to pursue additional certification as court interpreters, which requires passing an examination in Slovenian legal legislation.

Poland (STPJM): A two-tiered examination system is implemented and designed to support professional development. Candidates must first pass an initial practical interpreting exam conducted by a committee, granting them candidate status within the Association. Following a two-year candidacy period during which candidates are encouraged to actively participate in association activities, they become eligible to sit for a second examination that confers full membership status. This progressive system ensures both technical competency and meaningful integration into the professional community, treating the candidacy period not merely as preparation but as an opportunity for genuine contribution to the field.

Membership requirements for deaf and hearing interpreters are the same for thirty of the thirty-seven associations. Four associations report different membership criteria for deaf interpreters:

- Austria: General membership requirements remain consistent for both deaf and hearing interpreters, with the admission examination being slightly modified to reflect the specific work context of deaf interpreters.
- Belgium - Flanders: Hearing interpreters are required to hold a university degree in interpreting, while deaf interpreters must possess an interpreting certificate.
- Belgium - Wallonia: A different accreditation system applies for deaf and hearing interpreters.
- Switzerland - German region: Due to the absence of a permanent study program for deaf interpreters equivalent to that available for hearing interpreters, membership criteria historically relied on practical field experience. In recent years, studies for deaf translators have been established, and members now qualify either through obtaining this certificate or demonstrating relevant experience.

The other associations that report on deaf interpreter membership provide some additional information. For example, France does not currently offer training or a master's degree program for deaf Interpreters and, consequently, deaf Interpreters are not eligible for membership in AFTILS. However, deaf translators who hold a master's degree in translation may become members under the same requirements as hearing interpreters. While deaf Interpreters are not recognized within AFTILS, they have formed their own association to represent their professional interests. Additionally, deaf intermediators, for whom a dedicated master's degree program exists, have established their own professional association, and AFTILS maintains a close working relationship with this organization.

Poland evaluates both groups through the same qualification committee, adapting tasks to professional contexts. Deaf interpreters work from written text or International Sign video, while hearing interpreters work from spoken Polish. Some deaf interpreters note that equal continuous professional development (CPD) requirements feel disproportionate given fewer paid opportunities, though the system recognizes community support, advocacy, and allyship work where Deaf members often excel.

Scotland, Lithuania, Bulgaria, and Denmark apply unified membership standards. All interpreters must meet the same qualification, registration, and CPD requirements regardless of hearing status.

Ireland's CISLI recognizes multiple pathways to membership for both deaf and hearing interpreters, including third-level qualifications, historical accreditation processes, Trinity College's CPD Certificate in Deaf Interpreting, and RISLI registration.

Germany allows deaf interpreters to take the State exam and become members, though university-level interpreter training remains inaccessible to deaf students. Austria maintains consistent requirements with slightly adjusted admission examinations for deaf interpreters.

Belgium shows regional variation: Flanders requires university degrees for hearing interpreters but certificates for deaf interpreters, while Wallonia operates a different accreditation system. Switzerland's German region originally required field experience due to absent training programs but now accepts either certificates from recently established deaf translator programs or demonstrated experience.

Slovenia does not formally differentiate membership requirements but reports practical challenges. Deaf clients often prefer not to hire deaf interpreters, seeing no value in paying for two interpreters, and reportedly trust hearing interpreters more. Hearing interpreters increasingly resist support roles due to experiences of subordination and underestimation by Deaf interpreters.

Malta currently has no deaf interpreters, though its statute allows for future membership should they emerge.

Although the number of interpreter associations with deaf members has increased since 2020, it remains harder for deaf or hard of hearing persons to join their national interpreter association. The main reason is the requirement to have a degree from a national sign language interpreter training program, which are predominantly designed for hearing interpreters (see Part 3 - Deaf Interpreters).

1.10 Membership benefits

Figure 1.8 shows the benefits individual members receive when they are a member of the interpreter association.

Fig. 1.8 Individual membership benefits - number of associations per category (n=37)

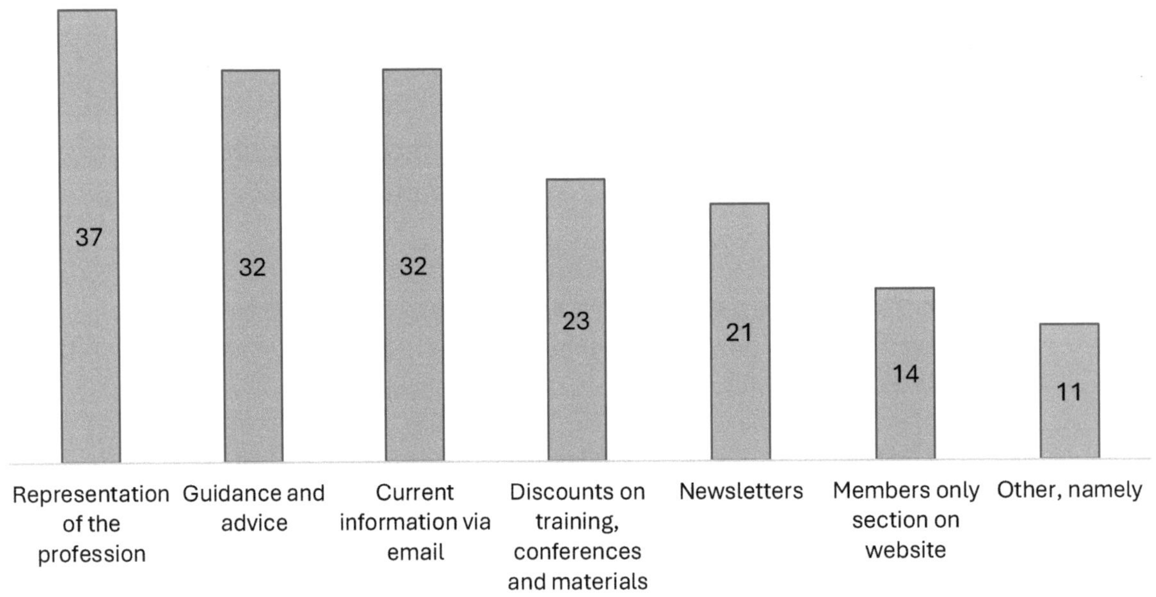

Professional recognition and credibility emerge as fundamental benefits in several countries. In Poland, where no national certification system exists, membership in the Polish Sign Language Interpreters Association (STPJM) serves as a widely recognized marker of quality and ethical practice:

> The deaf community and many institutions view STPJM membership as a guarantee of interpreting standards, with some public institutions even awarding additional points in procurement procedures to member interpreters. The association provides official confirmation letters for such purposes and maintains a "Find an Interpreter" directory where members can showcase their professional profiles, effectively serving as a quality assurance mechanism in the absence of state regulation.

Training and professional development opportunities feature prominently among membership benefits. The Finnish association offer collective agreements, training events, annual publications, and website articles to support interpreter growth. The Danish association provides access to training, conferences, and groups for personal reflection and supervision arranged by the association itself. AFTILS in France offers online conferences and discounts for specialized interpreter training programs, demonstrating how associations bridge gaps in continuing education infrastructure.

Communication networks and peer support have become increasingly vital, particularly through digital platforms. Ireland reports about "very lively WhatsApp groups" where members constantly share information, including a general group, a technology group established during the COVID19 pandemic, a group for theatre and arts interpreting, and even a private group for interpreters who wear hearing aids. Scotland similarly relies predominantly on WhatsApp for sharing information and providing support and advice. These informal digital networks complement formal communication channels and create communities of practice that extend beyond occasional meetings.

Practical professional support takes various forms across associations. Slovenia offers additional training, job search assistance, financial accounting services, and representation before authorities in cases of non-payment, and maintains records of all services while connecting interpreters to a nationwide network. ASLI (England, W&NI) provides insurance, wellbeing support, and networks for peer support, addressing both the business and emotional dimensions of interpreting work. Poland's job offers mailing list connects members with potential assignments without the association acting as a market intermediary, preserving interpreters' independence while facilitating access to opportunities.

Some associations provide opportunities for direct professional engagement. Serbia's Center offers members the chance to be professionally engaged through the organization itself. Belgium's Flemish association organizes informal gatherings and activities that strengthen community bonds. France provides access to legal assistance alongside networking opportunities, recognizing that interpreters face complex professional challenges requiring specialized support.

1.11 Aims of the association

The type of aims of the interpreter associations are about the same as in 2020 (Fig. 1.9). However, in 2025 more associations focus on offering training and continuing education, which came in fourth place in 2020. The aim of providing a network of support stays in first place. The other aims stay in the same place of priority as in 2020.

Fig. 1.9 Interpreter associations' aims - number of associations per category (n=37)

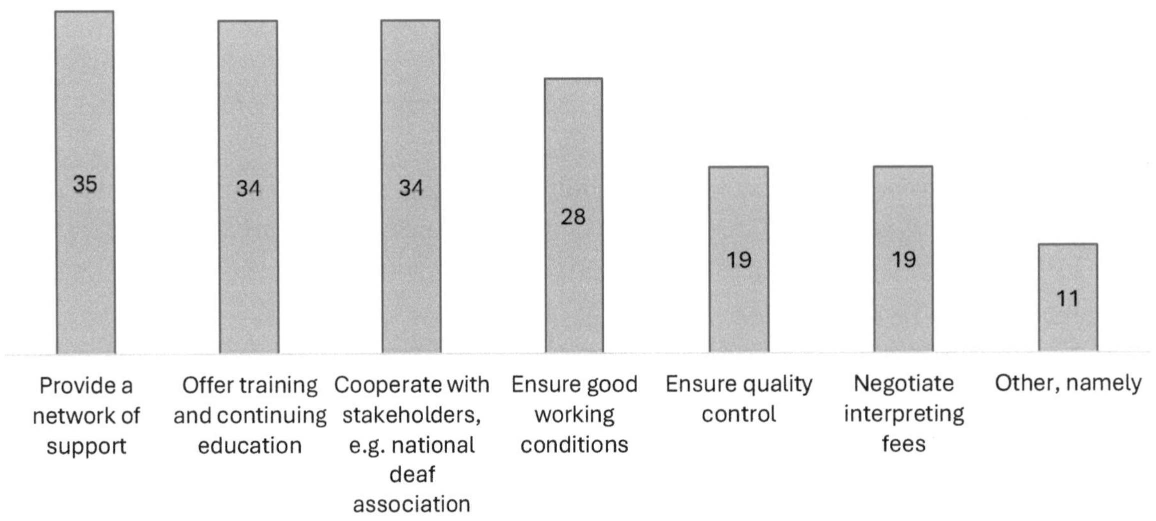

Other aims mentioned:

- Government advocacy and policy engagement: Discussing accessibility, professional conditions, and legal frameworks with government authorities (Portugal, Czech Republic, Slovenia, Sweden).
- Professional standards and quality assurance: Ensuring quality control and establishing professional working guidelines (Hungary, England, W&NI-ASLI).
- Working conditions and compensation: Advocating for fair pricing, good working environments, and professional conditions (Portugal, Czech Republic, Slovenia).
- Professional support and resources: Providing insurance, wellbeing support, business and legal advice to members (England, W&NI-ASLI, Netherlands).
- Deaf interpreter opportunities: Improving access to training and employment for Deaf interpreters (Poland).
- Sustaining the profession: Supporting interpreters to remain in the field and ensuring continued service to the national deaf community (Serbia).
- Knowledge sharing and professional development: Keeping members informed about current developments, international associations (WASLI, EFSLI, WFD, EUD), and training opportunities (Belgium Flanders).
- Professional promotion and evolution: Communicating about the profession to attract new generations and monitoring social changes to evolve the profession (France).

1.12 Collaboration with the deaf organization

One of the most important aims of the interpreter associations is collaboration with deaf associations. Nearly ninety percent (89,19%) of the national interpreter associations consider close collaboration with the national deaf association essential. They do this in various ways, most importantly in supporting each other's advocacy work with the government. The associations mention various forms of collaboration such as via a memorandum of understanding, working groups, and joint projects.

1.13 Funding of the association

Fig. 1.10 Funding of the interpreter association (n=37)

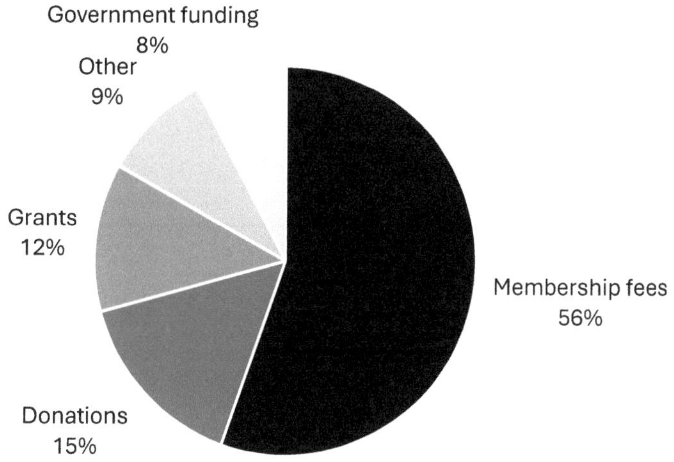

The most important source of funding of the interpreter associations are membership fees (Fig. 1.11). Other forms of funding are an e-shop (Czech Republic), conferences (Greece), and projects (Lithuania). The Polish association has the status of a Public Benefit Organization (OPP). This means that taxpayers in Poland can allocate 1.5% of their annual income tax to one chosen public benefit organization. The association receives such support from taxpayers indirectly, through the tax office, which distributes these contributions.

1.14 Membership fees

There are vast differences between associations' membership fees (Fig. 1.11). Calculating the total number of individual members (Fig. 1.3) and the individual membership fees (Fig. 1.11) provides the total amount that each association receives on individual interpreter membership fees (Fig. 1.12).

Fig. 1.11 National associations - 2025
Individual interpreter membership fees (in Euros)

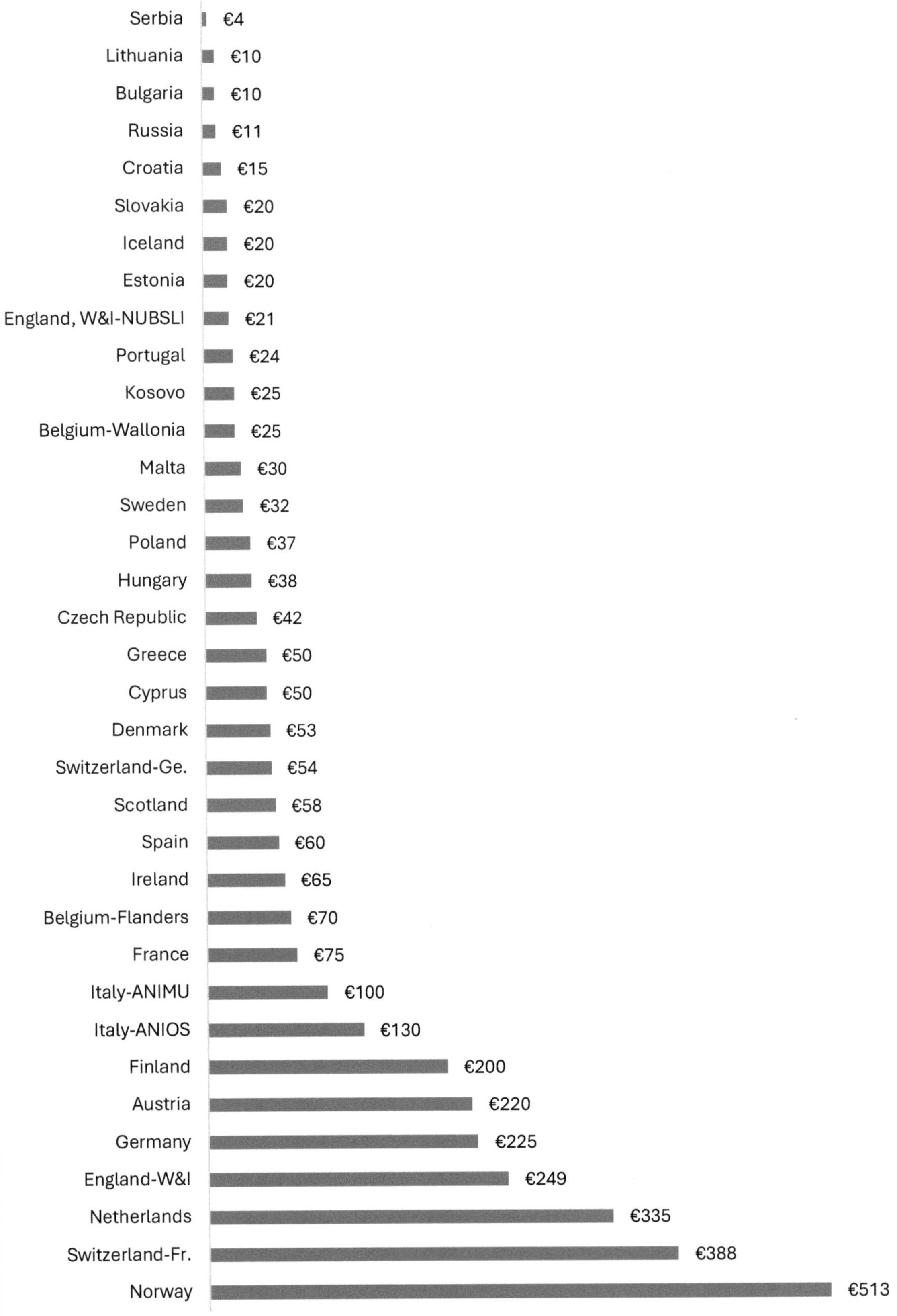

Fig. 1.12 Annual income per association based on individual membership fees (in euros)

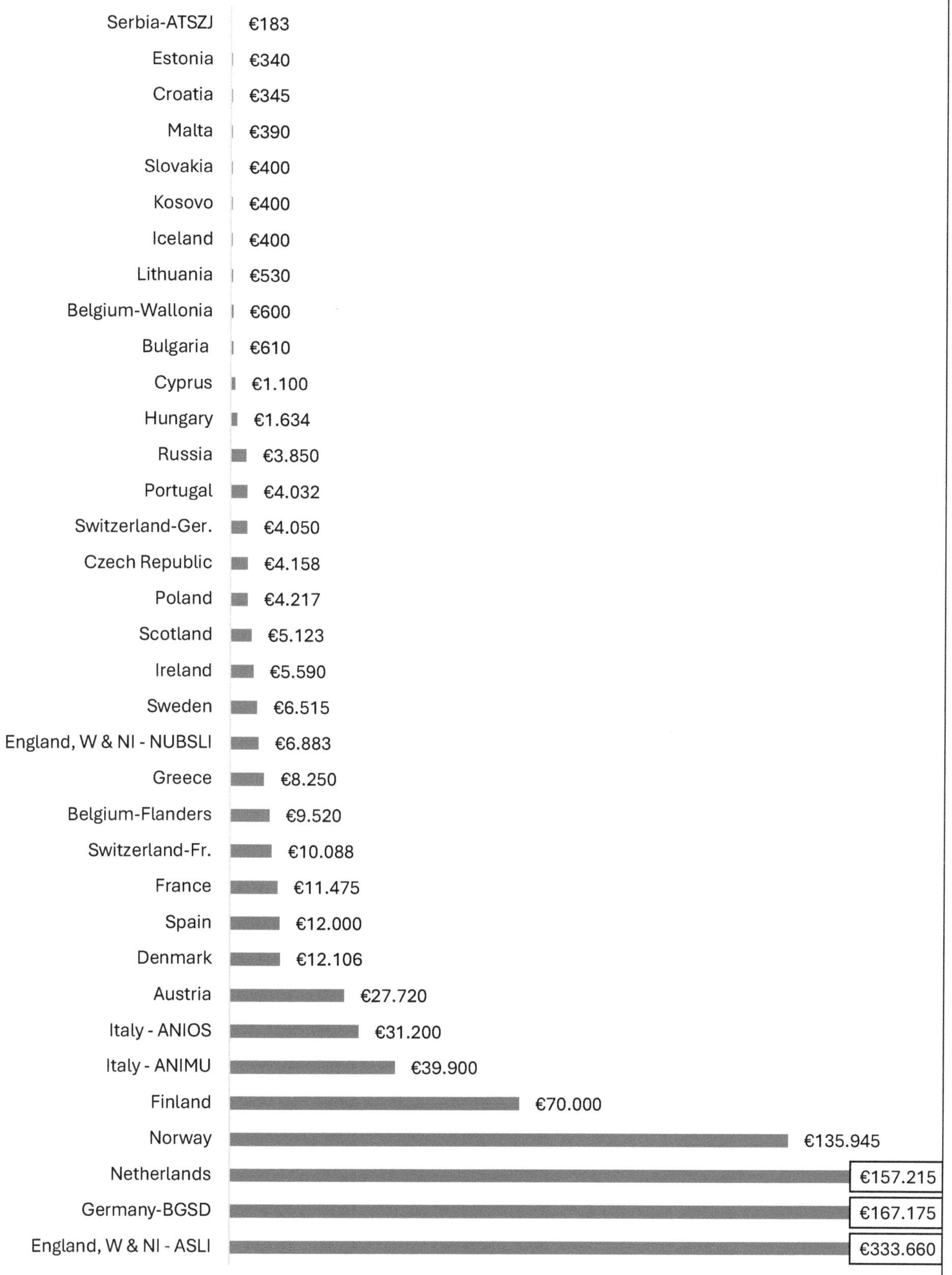

1.15 Associations' volunteer and paid positions

The thirty-seven national associations indicated their positions in their associations and if they are paid or not (Table 1.2). Most of the associations rely on volunteers, although there are a few paid staff members.

Table 1.2 Number of associations and their volunteer and paid positions (n=37)

Position	Volunteer	Paid
Board	36	0
President/chair of the association	29	2
Working groups and/or committees	29	0
Ethical/disciplinary/complaints committee	14	1
Advisory board	11	0
Chief financial officer	8	4
Other	8	1
Office staff	7	8
Daily manager/executive director/CEO	2	3

Associations appear to struggle with their workload, which is mainly carried out by volunteers. CISLI (Ireland) reports that they currently have no chairperson, vice chairperson and member care liaison, and if these positions are not filled by the next AGM, CISLI must close down. STPJM (Poland) mentions that according to Polish law, their Association is required to use professional accounting services. This involves a regular monthly payment to an external accounting company, as well as an additional annual expense for preparing the financial report. When STPJM organizes events that are interpreted into Polish or Polish Sign Language (PJM), they strive to pay interpreters from the Association's membership fees, unless they explicitly choose to waive their fee as a form of support for the Association. However, they make sure not to suggest or pressure interpreters to do so.

1.15 Other types of membership

Next to individual membership, the associations also indicate that they have other membership categories (Fig. 1.13). In the other category the associations mention: honorary and supporting members, partners and supporters, and retirees. In addition, there are some associations who do not have any other membership categories (Belgium-ABILS, Bulgaria, Hungary, Iceland, Serbia, Switzerland-ARILS).

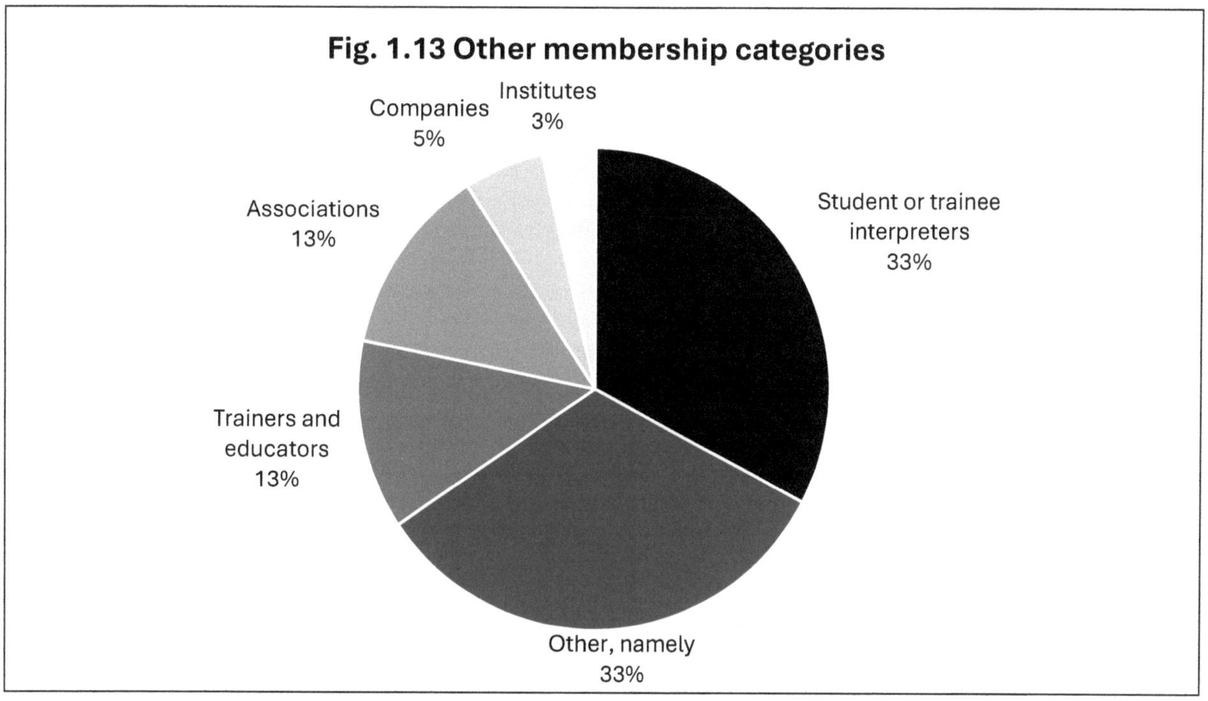

1.16 Main issues the associations are facing

The main issues the associations face are:

- Lack of funding and unstable financing
- Shortage of qualified interpreters and training pathways
- Weak professional recognition and regulation
- Poor working conditions and low fees
- Limited member engagement and volunteer capacity
- Need for stronger quality assurance, CPD, and support structures

Table 1.3 Main issues national associations of sign language interpreters report to face

Country / region	Association	Issues
Austria	ÖGSDV	Budget cuts in important sectors, not enough interpreters to meet the demand.
Belgium – Flanders	BVTG	No funding or grants — No quality control — not enough CPD Working conditions are still not obtained/still lacking.
Belgium – Wallonia	ABILS	Not enough resources (money) and not enough engagement from members.
Bulgaria	НАПЖЕБ	Shortage of qualified interpreters: - few professionals who are available due to personal/work commitments - uneven regional access - no backup coverage. Insufficient funding and support: - the work of the association often depends on sponsorships or project funds, if any, not stable budgets.

Table 1.3 Main issues national associations of sign language interpreters report to face

Country / region	Association	Issues
		Training gaps: - lack of enough mentoring opportunities. - insufficient CPD training available Implementation challenges. Public institutions often fail to provide interpreters despite legal obligations: - cumbersome service-request procedures. - limited access in health, education, and emergency settings. Low public and institutional awareness: - limited understanding of sign language and the interpreter's role. - undervaluing of professional standards.
Croatia	HDTPZJG	The new Law on Personal Assistance included SLIs, but this remains an ongoing struggle in Croatia. We have faced a year-long debate about the proper terminology, and both the Law on Croatian Sign Language and the Law on Personal Assistance describe SLIs as communication providers. This makes it difficult to lobby for and advocate for university-level programs.
Cyprus	ΣΔΚΝΓ	Our main issue is the certification and the lack of interpreters.
Czech Republic	CKTZJ	There is no one deaf organization; CKTZJ cooperates with various organizations and institutions based on their needs and our needs or goals.
Denmark	FTT	Focus on ensuring good work conditions and advocating for interpreting fees to increase.
England, W. & NI	ASLI	threat of AI generated sign language; inconsistent professional standards; real terms decline in numbers.
	NUBSLI	Interpreter Fee Guidance. Undercutting prices due to conflicting contracts in the areas, e.g.: NHS and local health care systems. Using interpreters' image for A.I consumption. Cuts to Access To Work/ Green Paper Consultation using our ATW Action group.
Estonia	EVK	No interpreter training now. Had the 3-year training program in the university in 2006-2019 and no training after that.
Finland	Kieliasiantun tijat	We try to improve the status of deaf interpreters in the interpreting services so that they would have the same opportunities to work as hearing SLIs (now restricted). We're constantly working to improve the working conditions as well as the fees of interpreters which bidding processes tend to weaken. We also comment on the procurement documents of interpreting services so the perspective of interpreters would be included in the process. Also, we're lobbying for a national legislation that would define the qualifications of all interpreters working in public administration or in the services offered by the public services. In connection with the legislation we're lobbying for a register that would cover all interpreters' whether they are SLI's or interpreting between spoken languages.
France	AFTILS	The development of AI: we still are figuring out our position. VRS are supposed to be open 24/7 in 2026, but the government does not seem to give solutions to strengthen the number of interpreters (the current political situation in France might be the main reason).
Germany	BGSD	The association was not very active in the past. a re-activation in many areas was the focus in our first year of the new board. We re-established contact with associations that had left the umbrella association, reactivated working groups, and offered additional webinars and training sessions. a lot more contact and conversations in general. Setting up the feedback or complaint mechanism has now priority, along with aligning with other associations (interpreters and general freelancers) concerning lobbying for or against new laws by the new government (in place since March).
Greece	SDENG	The main issues our association is currently facing are the official recognition of the profession by the state and its integration into university-level education.

Table 1.3 Main issues national associations of sign language interpreters report to face

Country / region	Association	Issues
Hungary	JOSZ	The salaries of interpreters are very low, and the salaries of freelancers have remained unchanged since 2009. This is why many people leave the profession. And it is difficult to train new recruits because the profession is not attractive to young people or other career changers.
Iceland	HART	No formal collaboration with other organizations.
Ireland	CISLI	We have no chairperson or vice chair and most of the new committee are inexperienced. One of the important meetings we attend (about the ISL Act) is only on in the afternoons. It is being run and attended by professionals who will not work after 5pm. It can be very hard for us to attend voluntarily during our working day, so we have decided to pay a member to represent us. A lot of work was done by the previous committee. They compiled a strategy for the next five years, a handbook for the committee, a CISLI charter and just completed a fees survey https://cisli.ie/resources/fees-surveys/ which is on the website. It feels like we need a better system of enticing members to the board, supporting current members and handing over to the new committee.
Italy	ANIMU	The proliferation of Sign Language associations within the same country and the fact that the regulation for becoming SL is not yet standardized, leading to a lot of organizations providing training that are not really adequate for the profession.
	ANIOS	The main activities involve discussions with state representatives on matters concerning the profession of sign language interpreters and their rights.
Kosovo	KASLI	Advocacy on providing the interpreter training
Lithuania	LGKVA	Lack of cooperation with the deaf community, raising the prestige of interpreters, certification issues, funding
Malta	AILSM	Funding, lobbying for another cohort of sign language interpreting course
Netherlands	NBTG	- Ensure quality control - Provide a support network - Offer high-level training and education, but we also want to offer interpreters various learning methods (webinars, podcasts, scientific articles) within a community via a learning platform, so that lifelong learning becomes a given. - Make NGT interpreter training accessible for Deaf interpreters - Make the agency (interpreter contact) more accessible/workable for Deaf and hearing interpreters. - Make the interpreting profession more sustainable because of the high drop-out rate and the high number of part-time workers - Negotiate interpreting fees
Norway	TiA	Freelance SL interpreting is subject to a tendering process. The process takes much longer than anticipated, and the tendering process does not take the user organizations or the unions sufficiently in account (in our opinion). General issues with proper pay for freelance interpreting.
Poland	STPJM	One of our main challenges remains the lack of national certification for sign language interpreters. We try to take part in governmental working groups and consultations, but this requires many hours of voluntary work and coordination. Since we do not have any paid staff, all administrative, organizational, and representational tasks are carried out by the Board and active members on a voluntary basis. While this is possible to manage, it demands continuous communication and coordination, which can be difficult when members' professional interpreting work naturally takes priority over association activities. Another recurring challenge is cooperation with our accounting firm, although the cost is low, obtaining the necessary documentation or information often requires a great deal of persistence and time, which adds to the overall workload and drains our limited capacity.
Portugal	ANAPI-LG	The lack of government responses regarding the regulation of the profession; the absence of standardization of interpretation service fees; situations of lack of professional ethics.
Russia	RASLI	The payment of the interpreters' work.

Table 1.3 Main issues national associations of sign language interpreters report to face

Country / region	Association	Issues
Scotland	SASLI	Overall, the association is good, though we have recently struggled with securing volunteers for the Board. Thankfully, we are currently at five members but having another one or two would be great.
Serbia	ATSZJ	Our Association faces significant challenges due to the lack of formal recognition, training, and professional opportunities for sign language interpreters in Serbia. There is no national interpreter training program, no university-level education in the field, and Serbian Sign Language (SSL) is still not documented, standardized, or taught as a linguistic discipline. Although the Constitution guarantees access to information in sign language, interpreters remain largely self-taught and community-trained, while the number of people entering the profession continues to decline. Although Serbia has adopted a Law on the Use of Sign Language and an Anti-Discrimination Law and has ratified the Convention on the Rights of Persons with Disabilities (CRPD), there have been no substantial systemic improvements. Our primary focus is to support the interpreters who remain, ensuring the Deaf community is not left without professional services. We strive to maintain quality through feedback from Deaf and hard-of-hearing individuals, and by offering pro bono workshops, mentoring, and shared resources. Our Association does not have the capacity for lobbying or advocacy, as members work voluntarily, and interpreters are often more committed to supporting the Deaf community in advocating for their rights than to advocating for the rights of interpreters themselves.
Slovakia	SKTPJ	The main challenge, both in the past and at present, lies in interpersonal conflicts and disagreements arising (even within the Board) from a division within the interpreting community. Broadly speaking, there appear to be two groups of interpreters: one representing a more "older" mindset (predominantly CODAs), and another comprising a "newer" generation of interpreters who are more open to continued education, feedback, and professionalization.
Slovenia	ZTSZJ	Reorganization of the association, change in legislation, professional training of interpreters, supervision of interpreters.
Spain	FILSE	Demotivation and lack of involvement of members, without external funding it is very difficult to maintain it, and lack of official quality training and precarious working conditions.
Sweden	STTF	Dwindling membership, extremely difficult to find people to join the board, harsher societal climate which affects terms and conditions for our colleagues and makes them less willing to be active and vocal in STTF and our work
Switzerland – Fr.	ARILS	We are facing a lack of enthusiasm to organize continuous training, as well as a loss of motivation among some members.
Switzerland – It.	ILISSI	Motivation, many members have not much motivation in professional development. No time, Family, etc.

Part 2 – Quality assurance

2.1 Quality assurance

Not every country has a national association of sign language interpreters. As mentioned in the introduction, if there is not a representative of a national association, the national deaf association or an individual expert has provided the responses to this survey. This chapter on quality assurance first covers replies from the national associations of interpreters. In the following sections (2.2 and onwards) all other topics related to quality assurance are presented and all respondents (N=47) are included.

In 1.11 the associations listed all their aims, of which one was 'to ensure quality control.' In a separate question, the associations were asked if their association actually focuses on quality assurance (Fig. 2.1).

Fig. 2.1 Does your interpreter association focus on quality assurance? (n=37)

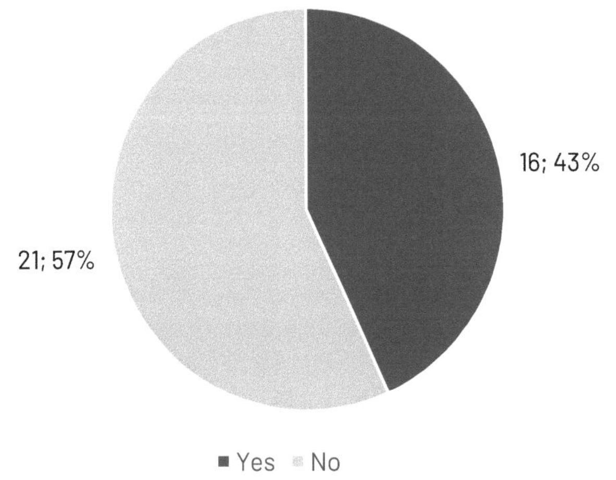

The national associations who answered yes (n=16) execute their quality assurance in various forms. For example, Bulgaria and France use compliance to the code of ethics. However, France does state they are unable to remove any members if they are not in compliance as they do not have a monitoring structure. Bulgaria states they also monitor the quality of interpreters by sending periodically emails focusing on the important parameters and ethics of the interpreter, providing individual guidance when necessary, and organizing seminars and Continuing Professional Development (CPD).

CPD as quality assurance in the form of seminars, trainings and conferences is organized by other countries such as Greece, Cyprus, France, Netherlands and Poland. Italy-ANIOS is required by law to annually check if their interpreter members have obtained a required number of educational credits to maintain their qualification.

Greece has a policy in place for more demanding interpreting assignments, where a newer member works alongside a more experienced interpreter to help improve the quality of the service provided: "This approach ensures both professional development and the maintenance of high standards in our field."

Every three years the Austrian association requires individual members to provide: (1) confirmation that they worked at least thirty hours per year, (2) participation in at least thirty hours of seminars or further training related to interpreting or translation, (3) attendance of at least one general meeting, (4) attendance of at least three working group meetings per year, (5) attendance of the ÖGSDV-seminar "Power and Responsibility" (every 6 years).

The Polish Association ensures quality control through three main mechanisms: (1) Qualification committees that assess interpreters applying for candidate and full membership. (2) A CPD system that requires members to earn a set number of points each year; members who remain inactive for more than three years lose their membership. (3) An ethics committee that reviews complaints and addresses issues related to professional conduct and ethical standards.

Next to being the representative body of sign language interpreters the Estonian association also provides the qualification of interpreters by running exams for interpreter candidates. Malta is working toward a professional warrant to be granted and renewed to interpreters only if they meet the quality criteria.

In the Netherlands they are working towards improving the quality of interpreting through stakeholder consultation (a consultative structure of professional associations for Dutch Sign Language (NGT) and sign language interpreters and advocacy groups for deaf and hard of hearing people). Specifically, the independent registry of sign language interpreters (RTGS) is exploring the possibility of a new assessment and a personal development profile for interpreters. Under discussion are also the rules of the independent complaints committee and the extent to which it has the authority to impose sanctions for violations of the professional code. The NBTG is also working with the registry to implement a specialization system, allowing interpreters to specialize in a specific field.

In Kosovo, Serbia, Slovenia and Russia quality assurance is achieved by close collaboration and consultation with representative associations of deaf and hard of hearing people. In Slovenia, if a negative report is received about an interpreter, they will be checked for qualifications. In Russia deaf persons are consulted about their satisfaction with the interpreting services through a continuous survey. In Serbia they address critical feedback by: "trying to improve quality through pro bono online workshops, mentoring sessions, and by sharing relevant readings, as we currently lack continuous training programs and financial resources to organize them."

Italy-ANIMU offers a dedicated section on their website to report any issues related to sign language interpreting services, this is for members and users of the services. Engl. W&NI-NUBSLI represent members who may be struggling in their career- late payments, contracts in the local area, undercutting of fees. They can provide legal advice through the Unite's Legal Team.

2.2 Definition in national legislation

To protect and regulate the profession, national legislations can include a definition of the professional sign language interpreter. Of the forty-seven respondents, twenty-two report having a definition of the profession in their national legislations and twenty-five do not. The provided definitions are in Table 2.1.

Table 2.1 Yes, definition of the sign language interpreter profession in the national legislation (n=22)	
Albania	Provisions from laws that mention interpreters Law No. 93/2014 "On the Inclusion and Accessibility of Persons with Disabilities" This law defines "interpreters or sign language interpreters and communication facilitators for persons with speech and hearing impairments." (Platforma PAK, qbz.gov.al) The law requires public institutions and service providers to take measures to make communication accessible, including the use of sign language and interpreters when necessary. (SMART Balkans) Criminal Procedure Code (CPC), Article 58, as amended It provides the right of the victim in a criminal proceeding to communicate in their own language and to have the assistance of a translator, sign language interpreter, or communication facilitator for persons with speech and hearing impairments.
Belgium-Flanders	https://www.vlaanderen.be/inter/toolbox-toegankelijke-steden-en-gemeenten/algemeen-bestuur-dienstverlening-en-communicatie/tolken-vlaamse-gebarentaal-vgt
Bulgaria	Art. 15. (1) (Amended and supplemented - SG, issue 10 of 2023, effective 31.01.2023) A translator from and into Bulgarian sign language is a capable natural person with Bulgarian citizenship or with the right of permanent residence in the Republic of Bulgaria, who possesses a professional qualification or a qualification in part of the profession "Translator from and into Bulgarian sign language" from the List of Professions for Vocational Education and Training under Art. 6, para. 1 of the Vocational Education and Training Act or has acquired a "Bachelor" educational and qualification degree in the specialty "Bulgarian Sign Language" from the professional field "Philology". (2) (Amended - SG, issue 10 of 2023, effective 31.01.2023) Translation services from and into Bulgarian sign language under the procedure of this chapter are performed by persons under para. 1, included in the List of Bulgarian Sign Language Interpreters and registered as persons exercising a freelance profession.
Croatia	The Law on Croatian Sign Language and Other Means of Communication for Deaf and Deafblind in the Republic of Croatia defines a 'communication intermediary,' which aligns with our profession rather than the term SLI. Additionally, the Law on Personal Assistance defines communication intermediaries and their work.
Denmark	You must have taken the national interpreting degree.
England W. &NI - NUBSLI	BSL Act 2022 outlines the details for BSL Interpreting in the UK.
Georgia	NA
Hungary	Sign language interpreter: a person who ensures the barrier-free transfer of information between deaf, hard of hearing, deaf-blind and hearing persons, interprets, translates, transcribes sign language texts from Hungarian sign language into Hungarian and from Hungarian into Hungarian sign language, performs written interpretation work, performs oral interpretation, and interprets, translates between other visual and tactile forms of communication using the special communication systems specified in the Annex, i) * deaf sign language interpreter: a person who is a native sign language user who ensures the barrier-free transfer of information between deaf, hard of hearing and deaf-blind persons, interprets, translates between sign languages or sign language codes, interprets and translates from and into other visual and tactile forms of communication, transcribes sign language texts, adapts written Hungarian texts into sign language, and, if necessary, works in cooperation with hearing sign language interpreters.

Table 2.1 Yes, definition of the sign language interpreter profession in the national legislation (n=22)	
Italy ANIMU	L'interprete di Lingua dei Segni Italiana (LIS) è un professionista specializzato che traduce e interpreta la comunicazione visivo-gestuale e verbale per garantire l'accessibilità a persone sorde o con deficit dell'udito. The Italian Sign Language (LIS) interpreter is a specialized professional who translates and interprets visual-gestural and verbal communication to ensure accessibility for deaf or hearing-impaired people.
Italy ANIOS	DECRETO-LEGGE 22 marzo 2021, n. 41 - The Republic recognizes the roles of the LIS interpreter and the LIST interpreter as professionals specialized in the translation and interpretation of LIS and LIST respectively, as well as in ensuring linguistic-communicative interaction between subjects who do not share their knowledge, through the translation into a codified visual-gestural modality of expressions used in verbal language or in other sign languages and tactile sign languages.
Kosovo	Sign language interpreter is the qualified person who has been trained and accredited by relevant institution.
Latvia	Disability Law. "Sign language interpreter - a person who ensures that a person with a hearing disability communicates with other persons in sign language, including in a vocational basic education, vocational secondary education or higher education institution where a person with a hearing disability is studying an educational program."
Lithuania	This is stated in official documents and legislation.
Moldova	1. Law No. 264 of 2008 (Justice Sector) The Law on the status, authorization and organization of the activity of interpreters and translators in the justice sector defines: Interpreter – a natural person specialized in oral translation from one language to another and/or in the translation of signs of the deaf, mute or deaf-mute, thus mediating understanding between two or more people. 2. Government Decision No. 333 of 2014 (Provision of Services) The Regulation on the provision of communication services through the use of sign language/Sign Language with the help of an interpreter, approved by this Decision, provides a more specific definition: Interpreter – a person specialized in communication through the use of sign language/sign language, thus ensuring understanding between the person/persons with hearing impairments and another person/persons within different authorities/institutions/organizations.
Netherlands	Interpreter: A translator of (spoken) text into (written) words or signs who is registered in the Register of Sign Language Interpreters and Sign Language Interpreters (RTGS External link: www.stichtingrtgs.nl). A communications assistant cannot act as an interpreter. https://wetten.overheid.nl/BWBR0049587/2024-04-19
Portugal	The profession is established in the Constitution, Law - 89/99.
Serbia	According to the Law on the Use of Sign Language (Zakon o upotrebi znakovnog jezika), a sign language interpreter is a person who has acquired the professional competencies to perform the duties of a sign language interpreter, that is, the competencies to translate spoken language into sign language and sign language into spoken language.

Table 2.1 Yes, definition of the sign language interpreter profession in the national legislation (n=22)	
Slovakia	In Act No. 448/2008 Coll. on Social Services, interpreting services are addressed along with the qualification requirements as follows: "Interpreting under this Act is carried out by a sign language interpreter, an articulatory interpreter, and an interpreter for deafblind persons in accordance with a specific regulation, unless otherwise provided in paragraph 11." 1. Interpreting in sign language may be carried out by a person who has completed at least full secondary general education or full secondary vocational education, and who has completed an accredited course in sign language interpreting of at least 350 hours, including a minimum of 280 hours of practical communication training. 2. A person who has completed higher education at the first or second level in the field of special education with a focus on surdopedagogy and has completed an accredited course in the relevant form of interpreting, including practical communication training as specified in points (a) to (c) for the respective form of interpreting. 3. A person who has completed higher education at the first level in the study program Slovak Language in Communication with the Deaf.
Slovenia	A sign language interpreter is a person who interprets Slovenian spoken language into sign language for deaf persons and sign language into Slovenian spoken language for hearing persons. A deaf person may exercise the rights under this Act for interpreting Slovenian spoken language into sign language and sign language into Slovenian spoken language. In areas where the Italian or Hungarian national community lives, a deaf person may also exercise the rights of interpreting Italian or Hungarian spoken language into Italian or Hungarian sign language and for hearing persons, interpreting Italian or Hungarian sign language into Italian or Hungarian. A sign language interpreter is an adult who has been issued a certificate in accordance with the regulations governing national professional qualifications and is entered in the register of sign language interpreters. Interpreters are obliged to comply with the code of professional ethics for interpreters in their work. The certificate is a public document. The provisions of this Act regarding sign language apply to court interpreters, and in other respects the regulations on courts apply.
Spain	Law 27/2007, of October 23, Article 2. Definitions. "A sign language interpreter is understood to be a person who, with the appropriate qualifications, performs interpretation between a sign language and an oral language, or vice versa, facilitating communication between deaf, hard-of-hearing or deafblind people, and hearing people."
Turkey	NA
Ukraine[HI]	Ensuring interlingual and intercultural communication between people with hearing impairments and those who can hear in various spheres of social activity.

2.3 Regulation of the sign language interpreter profession

The EU holds a database of all regulated professions.[19] There are different regulations, such as:

- Reserves of activities: activities are reserved to the holders of a specific professional qualification
- Protected title (without reserves of activities): only the title is protected
- Reserves of activities and protected title: reserved activities and protected title.

The respondents were asked whether the sign language profession is regulated in their country or region. The results are pictured in Figure 2.2. In most countries, the sign language interpreting profession is not regulated (52%).

Fig. 2.2 Is the sign language interpreter profession regulated in your country?

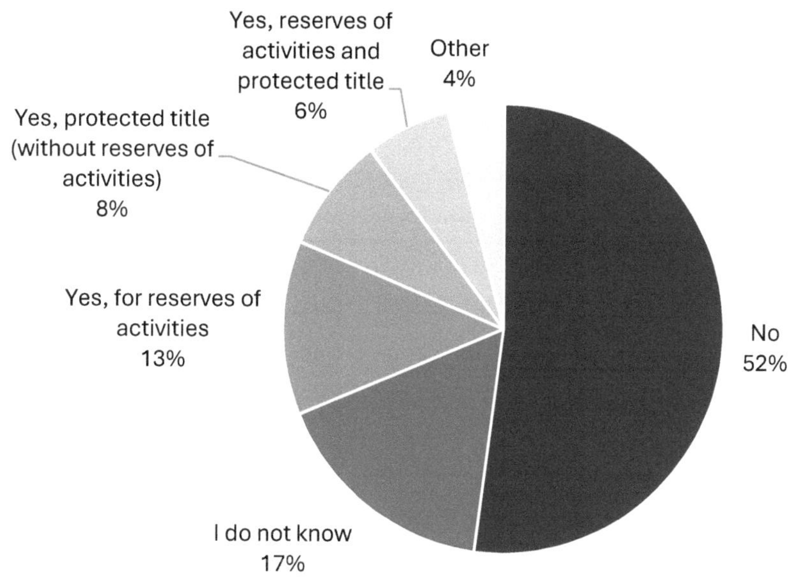

Serbia mentions that the profession of sign language interpreter is partially regulated: "The Law on the Use of Sign Language recognizes interpreters and defines their role, and the Ministry of Justice accredits interpreters for courts and police." In Kosovo the interpreter profession is close to being recognized by the National Qualification Agency.

[19] https://ec.europa.eu/growth/tools-databases/regprof/home

Fig. 2.3 Are there other professionals in your country/region who are put in place instead of qualified interpreters? (N=47)

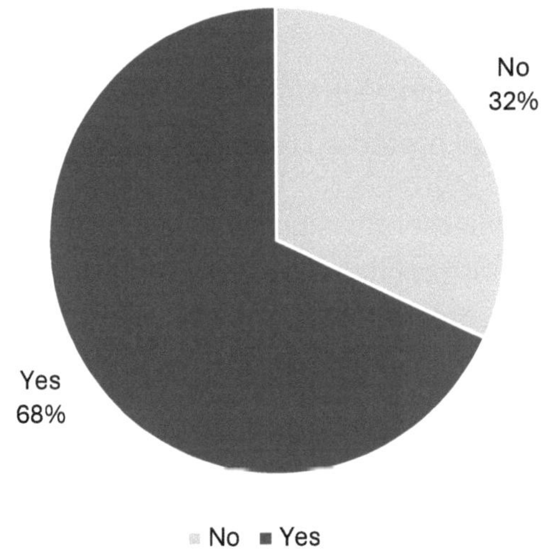

Out of forty-seven countries and regions, thirty-two reported that sometimes other professionals are employed instead of qualified sign language interpreters (Fig. 2.3). Professionals that can act as substitute interpreters include teachers, teaching assistants, social workers, or communication assistants. Such substitutions are reported to happen mainly in educational settings, as well as in sheltered employment, and church services.

2.4 Registry of interpreters

A registry or a registration authority is an organization which has the responsibility to register and maintain an official database of the qualified interpreters within a region or country. The registry can be used, for example, by consumers of interpreting services or public authorities. The registry usually has a set of criteria that qualifies an interpreter to register. The aim of the registry is not only to maintain a formal list of interpreters but also to provide a quality guarantee and ensure the qualifications of the interpreters who are registered. In 2012 the establishment of registries was a new European trend in the development of the sign language interpreting profession. In that year a total of fifteen countries reported to have a national registration authority, and in 2016 this number went down to twelve. In 2020 the number grew to seventeen countries, and to twenty-two in 2025 (Table 2.2).

Table 2.2 Country/region with an official SLI accreditation/registration body (n=22)		
Country / region	**Name official body**	**Independent or part of another association**
Albania	Albanian National Association of the Deaf (ANAD)	Part of the national deaf association
Belgium – Flanders	Flemish Agency for Persons with Disabilities (VAPH)	Independent
Bulgaria	Accreditation: The National Agency for Vocational Education and Training (NAVET) Register: The Agency for Persons with Disabilities (AHU)	Independent
Denmark[DI]	Den Nationale Tolkemyndighed	Independent
England, Wales & NI (ASLI, NUBSLI: 2)	National Registers of Communication Professionals working with Deaf and Deafblind People (NRCPD); Regulatory Body for Sign Language Interpreters and Translators (RBSLI)	Independent
Estonia	National Qualification Committee of the EVK	Part of the national association of sign language interpreters
Finland	Finnish National Agency for Education	Independent
Georgia		
Germany - BGSD	State Examination Office	Independent
Hungary	Government Office of the Capital City Budapest	Independent
Ireland	Register of Irish Sign Language Interpreters (RISLI)	Independent
Kosovo	Governmental register of sign language interpreters	Part of the national deaf association
Latvia	LAD Rehabilitation center	Part of the national deaf association
Malta	The Maltese Sign Language Council currently	Independent
Netherlands	Foundation Register of Sign Language Interpreters and Speech-to-Text Interpreters (RTGS)	Independent
Norway	The Norwegian National Register of Interpreters	Part of the directorate for integration and diversity
Romania	National Authority for Persons with Disabilities (ANPD)	Independent
Russia	Certification Center under the Ministry of Labor and Social Protection of the Russian Federation with the Russian Association of Sign Language Interpreters	Part of the national association of sign language interpreters
Scotland	The Scottish Register of Language Professionals with the Deaf Community (SRLPDC); National Registers of Communication Professionals working with Deaf and Deafblind	Independent
Serbia	The Ministry of Justice	Independent (Ministry)
Slovenia	Ministry of Labor, Family, Social Affairs and Equal Opportunities	Independent (Ministry)
Turkey	Turkish Sign Language Scientific and Approval Committee	Independent
Ukraine[HI]	Central Attestation Commission	Part of the national deaf association

Some countries and regions have more than one body that registers or accredits sign language interpreters: Germany (2), Russia (2), and Scotland and England Wales and NI (3). Most registration bodies are independent (n=14). The registration bodies in Estonia and Russia are part of the national interpreter associations. In Kosovo, Latvia and Ukraine the registration process is run by the deaf association and in Norway it is managed by the IMDi (the directorate for integration and diversity).

Table 2.3 Country/region *without* an official SLI accreditation/registration body (n=25)		
Austria	Greece	Portugal
Belgium - Wallonia	Iceland	Slovakia
Bosnia-Herzegovina	Italy (ANIMU, ANIOS: 2)	Spain
Croatia	Lithuania	Sweden
Cyprus	Luxembourg	Switzerland – Fr.
Czech Republic	Moldova	Switzerland – Ge.
Denmark	Montenegro	Switzerland – It.
France	Poland	

The individual deaf interpreter from Denmark reports:

> There is The National Interpreting Authority (Den Nationale Tolkemyndighed)[20]:
>
> "DNTM (The National Interpreting Authority) is a national agency under the Ministry of Social Affairs, Housing and Senior Citizens in Denmark. It administers and grants interpreting services for citizens with hearing disabilities. This primarily covers: Private life and leisure - e.g. medical appointments, leisure activities, associations, family events. Healthcare - e.g. hospitals, doctors, psychologists. Contact with public authorities - e.g. meetings with the municipality, state administration, courts. Education and employment are covered by other schemes and ministries (e.g. STUK under the Ministry of Children and Education, and job centers/the Danish Agency for Labor Market and Recruitment under the Ministry of Employment).
>
> Interpreting providers can only deliver services through DNTM if they are admitted to the system via a public tender. In total there were 309 interpreters registered (which is not compulsory) in 2023: 300 hearing sign language interpreters, of which nine are registered as interpreters for deafblind persons."

The number of interpreters per country or region that are registered in the registration bodies are listed in Figure 2.4. Germany and Belgium-Flanders are not listed in the figure as they indicate that they do not have an exact number. All registration bodies allow hearing interpreters to register. In three of the twenty-three countries with a registry, it is not possible for deaf interpreters to register: Kosovo, Latvia, and Serbia. Only five respondents have data on the exact number of deaf interpreters that are registered: Georgia (4 deaf and 2 hard of hearing), Ireland (24), Netherlands (5), Turkey (0), and Ukraine (4).

[20] The National Interpreting Authority.

Fig. 2.4 Registered interpreters in countries & regions with a registry

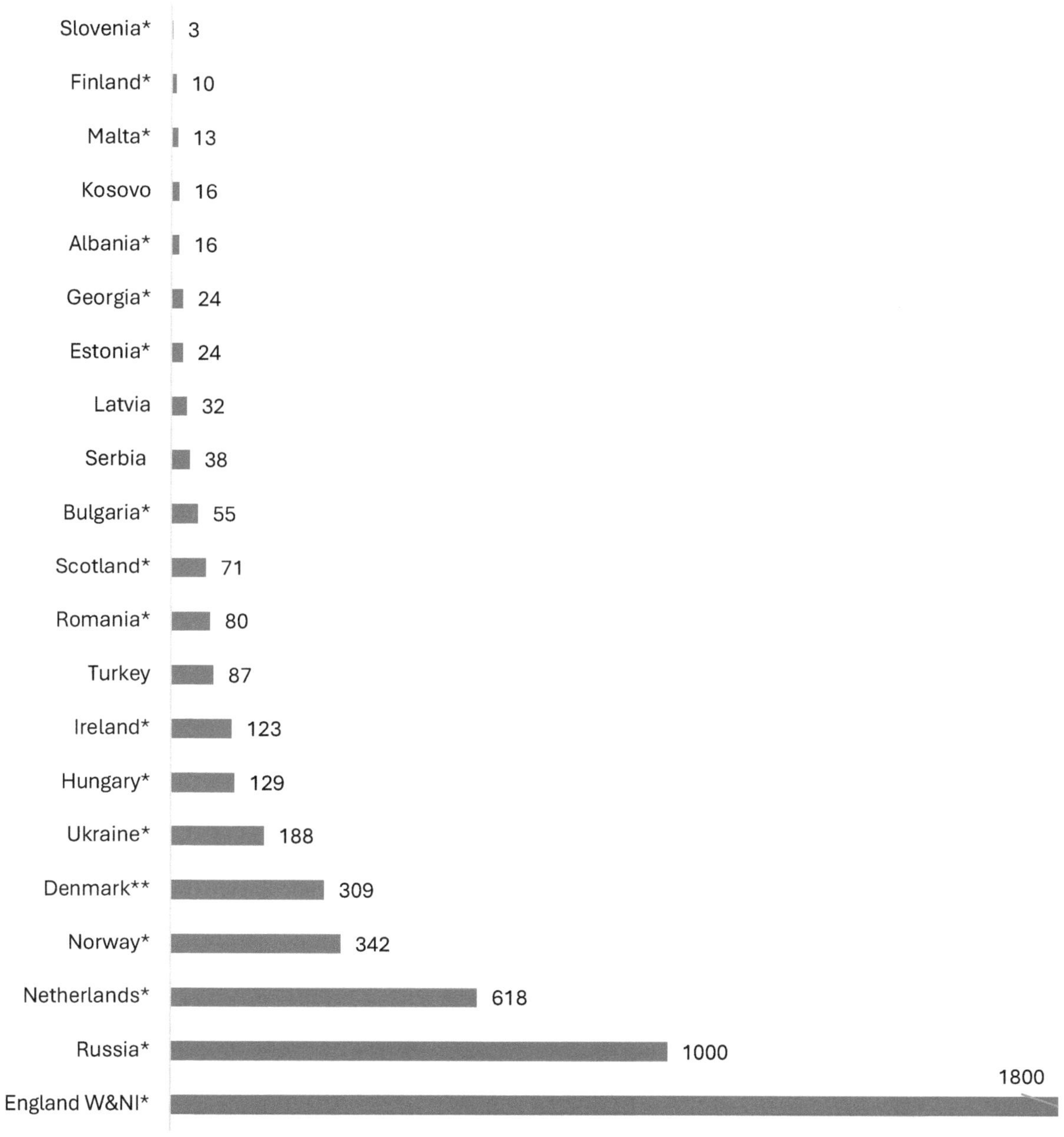

**Deaf interpreters can register*
***Response from Denmark [DI]: 309 in 2023*

The respondents were also asked how many registered interpreters interpret for deafblind persons. In Figure 2.5 an overview is provided of those countries that have provided the number of interpreters who are registered or accredited to work with deafblind persons.

Fig. 2.5 Number of interpreters accredited or registered to work with deafblind persons

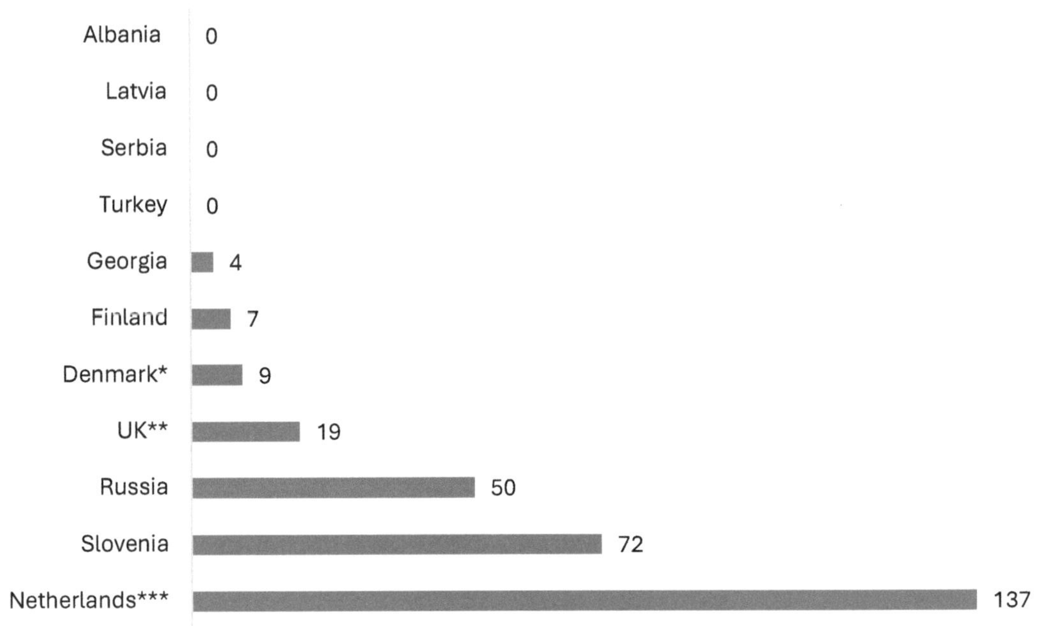

**https://www.dntm.dk/*
*** https://www.nrcpd.org.uk/registration-figures*
**** https://www.stichtingrtgs.nl/zoek-een-tolk (indicated by interpreters themselves)*

Table 2.4 presents an overview on the prerequisites per country to register as a sign language interpreter. Some registries have a website, find the links on: https://www.mayadewit.nl/european-study

Table 2.4 Overview of prerequisites and criteria per country to register as a sign language interpreter

Country / region	Prerequisites for registration	Is registration compulsory?	Comment
Albania	No fully established national registry; anticipated prerequisites include certification in Albanian Sign Language, recognized training, documented experience, and compliance with emerging standards.	No	Since 2023, limited inclusion in the Ministry of Justice National Registry for legal settings; system remains in early development.
Belgium-Flanders	Recognized interpreting qualification and adherence to a professional code of conduct.	Not compulsory (required for government-funded work)	Private practice without registration is possible but rarely sustainable; operates within an established ethical framework.

Table 2.4 Overview of prerequisites and criteria per country to register as a sign language interpreter

Country / region	Prerequisites for registration	Is registration compulsory?	Comment
Bulgaria	Approved training qualification; secondary education; linguistic competence; experience; medical fitness; ethical compliance; CPD commitment.	Yes	Registry maintained by the Agency for People with Disabilities; some practitioners remain unregistered.
Engl. W&NI - ASLI	Qualifications aligned with National Occupational Standards insurance Adherence; Code of Conduct Criminal Record disclosure.	No	Operates within professional regulatory structure.
Engl. W&NI-NUBSLI	Formal qualification pathway including supervised trainee status. Certificate of qualification.	Yes	Registration within organizational framework.
Estonia	Successful completion of the national qualification examination aligned with national qualification standard for sign language interpreting.	Required for most state-funded services	Certification overseen by multidisciplinary committee; deafblind training included but specialist practice limited.
Finland	Legal age and capacity; qualification in legal interpreting or higher education degree appropriate for the work of an interpreter and legal interpreting studies.	No	Register of Legal Interpreters provides formal recognition.
Georgia	Not specified.	Yes	Data not available.
Germany-BGSD	Formal education and/or documented professional experience supported by certificates.	No	Recognition linked to organizational membership; requirements vary by federal state.
Hungary	Clean criminal record; state-recognized qualification; ≥300 interpreting hours; CPD every three years.	Required for state-funded services	System expanding to include deaf interpreters once training is completed.
Ireland	NFQ Level 7–9 ISL/English interpreting qualifications or equivalent; possible to apply without qualifications but additional evidence needed.	No	Mandatory for public bodies; voluntary otherwise. General lack of recognition and provision of communication services for deafblind persons.
Kosovo	Completion of required training and accreditation process.	Yes	Registration follows accreditation.
Latvia	Higher education qualification and attestation determining professional category.	Yes	No formal Deafblind specialist provision.
Malta	Degree or higher qualification in sign language interpreting.	Yes	Specialist Deafblind provision under development.
Norway	Recognized interpreting degree.	No	Recently sign language interpreters incorporated into registry for spoken language interpreters; assignments mainly via NAV system and not through the registry.
Netherlands	Recognized diploma from Utrecht University of Applied Sciences.	Yes	RTGS registry developing specialist expertise pathways.
Romania	Romanian Sign Language course plus university degree.	Partially mandatory	Required with authorities and certain broadcasters.
Russia	No national examination requirement.	Yes	No unified compulsory national registration system.
Scotland	Professional qualifications; PVG membership; insurance; 30 CPD hours annually.	Yes	Structured regulatory framework; some interpreters registered via NRCPD.
Serbia	Registry at the Ministry of Justice registry: qualifications, ≥5 years' experience; clean record.	Yes	Required for court and police interpreting; Deaf interpreters not accredited in practice.

Table 2.4 Overview of prerequisites and criteria per country to register as a sign language interpreter

Country / region	Prerequisites for registration	Is registration compulsory?	Comment
Slovenia	National Vocational Qualification (NPK) examination.	Yes	Registration follows national vocational qualification.
Turkey	Ministry-coordinated examinations involving the Turkish Sign Language Scientific and Approval Committee.	Required for public institutions	No single centralized registry. Membership entry according to each association their own criteria.
Ukraine[HI]	Certification examinations every five years.	No	Certification authorizes qualification category. Certain category allows for legal interpreting.

2.4 Complaints' procedure

Of all the countries, twenty-five have a complaints' procedure in place. Table 2.5 shows all the countries and regions that have a complaints' procedure, including which stakeholders can file a complaint. Overall, most countries and regions do not have an official complaint mechanism in place. The countries that have a website for the complaint mechanism are listed here: https://www.mayadewit.nl/european-study.

Table 2.5 Countries with a complaints' procedure – parties that can file a complaint					
Country / region	**Anyone**	**Deaf / deafblind client reporting**	**Hearing client reporting**	**Interpreters (reporting on their colleagues)**	**Other, namely**
Albania	X	X		X	
Austria	X	X	X	X	
Belgium-Flanders	X	X	X	X	
Belgium-Wallonia		X	X	X	No official complaint mechanism
Bosnia and Herzegovina					No official complaint mechanism
Bulgaria	X	X		X	
Croatia					No official complaint mechanism
Cyprus					No official complaint mechanism
Czech Republic	X	X	X	X	
Denmark	X	X	X	X	
England, W. & NI		X	X	X	The employer or interpreting regulatory body
Estonia	X	X	X	X	No official complaint mechanism
Finland	X	X	X	X	The association of deaf/deafblind
France		X	X	X	No official complaint mechanism
Georgia	X	X	X	X	
Germany		X	X	X	No official complaint mechanism
Greece		X	X	X	
Hungary	X				No official complaint mechanism
Iceland	X	X	X	X	
Ireland					No official complaint mechanism
Italy-ANIMU		X	X		No official complaint mechanism
Italy-ANIOS	X	X	X		
Kosovo		X	X	X	No official complaint mechanism
Latvia		X	X	X	
Lithuania	X	X		X	
Luxembourg					No official complaint mechanism
Malta	X	X	X	X	
Moldova					No official complaint mechanism
Montenegro					No official complaint mechanism
Netherlands		X	X		
Norway		X	X	X	
Poland		X			
Portugal		X	X	X	No official complaint mechanism
Romania					No official complaint mechanism
Russia		X	X		
Scotland	X	X	X	X	
Serbia					No official complaint mechanism
Slovakia		X			No official complaint mechanism
Slovenia	X	X	X	X	
Spain					No official complaint mechanism
Sweden		X	X	X	No official complaint mechanism
Switzerland-Fr.	X	X	X		
Switzerland-Ge.		X	X	X	

Table 2.5 Countries with a complaints' procedure – parties that can file a complaint					
Country / region	**Anyone**	**Deaf / deafblind client reporting**	**Hearing client reporting**	**Interpreters (reporting on their colleagues)**	**Other, namely**
Switzerland-It.		X	X		
Turkey	X	X	X	X	
Ukraine[HI]		X	X	X	

Almost half of the countries and regions indicate that most of the deaf clients (Fig. 2.6) and some of the hearing clients (Fig. 2.7) are aware of the complaints' procedures.

Fig. 2.6 Are deaf clients aware of the complaints' procedure?

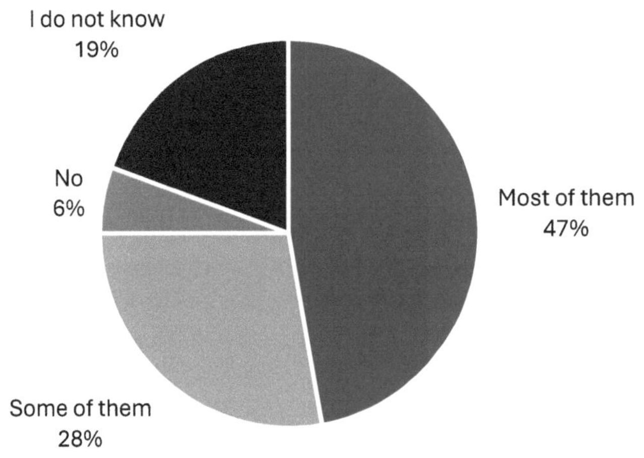

Fig. 2.7 Are hearing clients aware of the complaints' procedure?

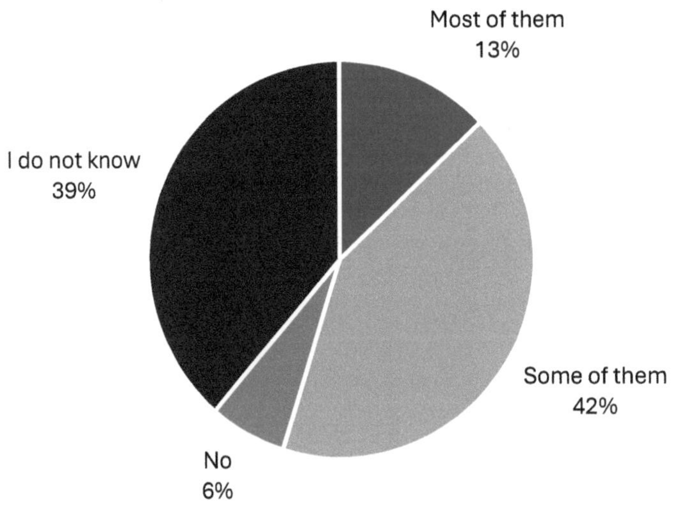

The respondents were also asked to indicate if the deaf and hearing clients use the possibility to file a complaint (Figures 2.8; 2.9).

Fig. 2.8 When deaf clients are not happy with the services received, do they use the possibility to file a complaint?

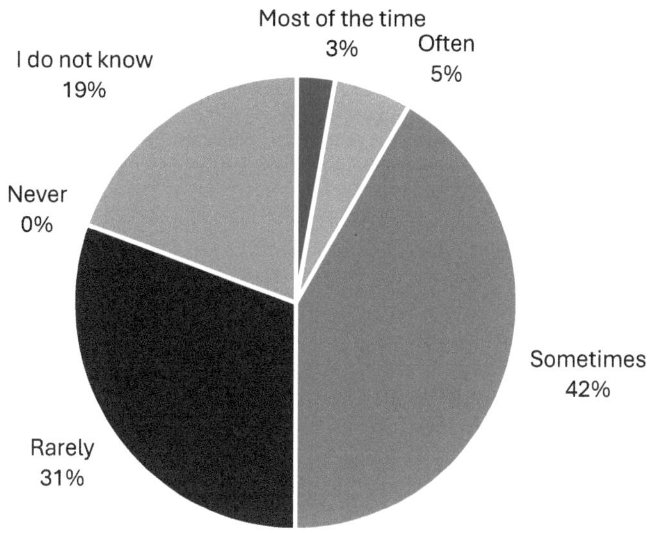

Fig. 2.9 When hearing clients are not happy with the services received, do they use the possibility to file a complaint?

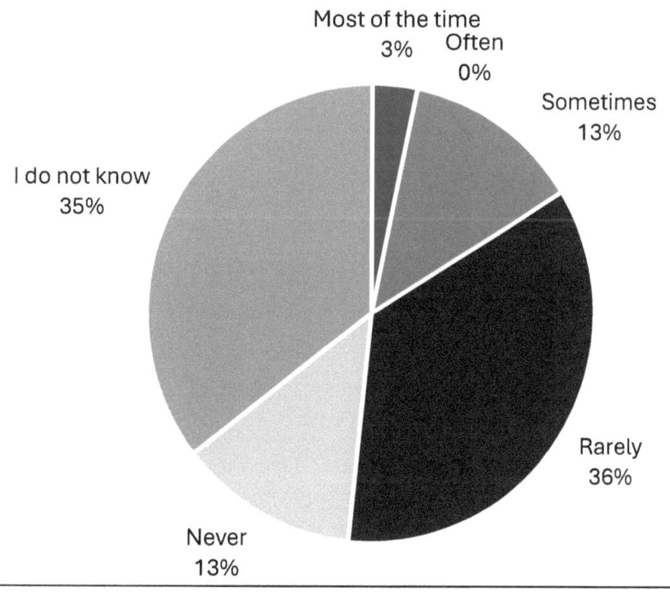

Of the countries without formal complaints procedures, countries state that instead of a formal complaint, the interpreter can receive feedback through the interpreting agency, personal feedback or informal, often face-to-face communications. Table 2.5 provides an overview of the alternative mechanisms through which interpreters receive client feedback.

Table 2.6 Alternative mechanisms through which interpreters receive client feedback

Country / region	Alternative feedback mechanisms
Belgium-Wallonia	Feedback provided via SLI services.
France	Clients may contact the interpreter directly or the interpreting service.
Germany	Feedback received via email, online postings, or direct communication.
Italy-ANIMU	Each association maintains a *'sportello del consumatore'* (consumer help desk) through which clients may submit complaints regarding services provided by members.
Kosovo	Feedback communicated via the interpreter coordinator responsible for liaising between the national association and government bodies.
Luxembourg	Individual interpreters may request feedback directly; at the institutional level, feedback may be managed through social services within the Ministry for Family Affairs.
Romania	Feedback provided directly by deaf clients to interpreters, without formal evaluation procedures or repercussions.
Serbia	Feedback provided verbally and directly, or through Deaf organizations. Informal community discussions and social media may also disseminate evaluative commentary.
Slovakia	No central national complaint mechanism. When interpreters are employed by regional organizations, clients may submit complaints to the employer, who is responsible for follow-up action.
Sweden	No national authority overseeing complaints. Governmental bodies may have internal procedures. For interpreters authorized by *Kammarkollegiet*, authorization may be reviewed, and sanctions such as formal warnings or revocation may be applied.

When interpreters receive a complaint, it can have consequences. Table 2.7 provides an overview of the different themes from the responses on the possible consequences following a complaint. In most countries the consequences are decided on a case-by-case basis.

For the countries that have a website for the complaint mechanism, please find the links on https://www.mayadewit.nl/european-study

Table 2.7 Consequences following a valid complaint (n=37)

Country / region	No consequence	Warning / Reprimand	Guidance / Discussion	Suspension	Expulsion /Removal	Loss of Qualification /Registration	Legal / Financial	Unclear/ Case by case
Albania	X							X
Austria			X	X				X
Belgium – Flanders	X							
Belgium – Wallonia								X
Bulgaria		X						
Czech Republic			X					X
Denmark	X							
Engl., W.&NI – ASLI				X	X	X		X
Engl., W.&NI – NUBSLI								X
Estonia						X		
Finland								X
Germany					X			X
Greece		X		X	X			
Iceland								X
Italy – ANIMU								X
Italy – ANIOS		X		X	X			
Kosovo								X
Latvia			X					X
Lithuania		X						
Luxembourg								X
Malta								X
Netherlands								X
Norway			X			X		X
Poland	X						X	
Portugal					X			X
Romania	X							
Russia		X					X	
Scotland						X		
Serbia							X	X
Slovakia								X
Slovenia		X			X	X	X	
Sweden	X							X
Switzerland – Fr.			X					X
Switzerland – Ge.	X		X					
Switzerland – It.			X					X
Turkey		X						
Ukraine[HI]		X						X

PART 3 – DEAF INTERPRETERS

3.1 Deaf interpreters – terminology

Each country uses different terms to refer to sign language interpreters who are deaf (Fig. 3.1). Some countries use more than one term. The most common term is 'deaf interpreter.'

France mentions that it depends on the activity:

- "Interprète sourd·e" ("Deaf interpreter" no matter the gender in French)
- "Intermédiateur·ice" ("intermediator" no matter the gender in French)
- Sometimes they add "Deaf intermediator", or
- "Traducteur·ice sourd·e" ("Deaf translator" no matter the gender in French)

Sweden mentions that the term 'sign language interpreter' is used to include all interpreters, deaf and hearing.

In the category 'Other' respondents mention the use of the following terms: relay interpreter, deaf translator, and deaf mediator. Slovakia says that because they do not have any deaf interpreters, they also have no term to refer to interpreters who are deaf.

In this publication the term 'deaf interpreter' is used throughout to mean sign language interpreters who are deaf.

Fig. 3.1 Term mostly used in the countries/region when referring to sign language interpreters who are deaf

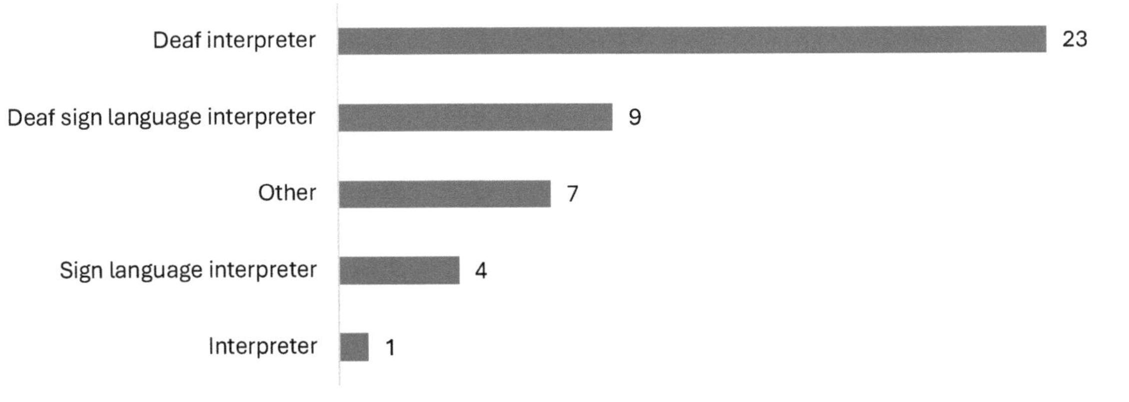

3.2 Deaf interpreter developments

This 2025 edition has for the first time a comprehensive overview of the developments of the deaf interpreters as professionals. Most of the respondents mention that the profession of deaf interpreters has improved in their country or region during the last five years or remained the same (Fig. 3.2).

Fig. 3.2 How has the profession of sign language interpreters developed for deaf interpreters in your country/region during the last five years? (N=47)

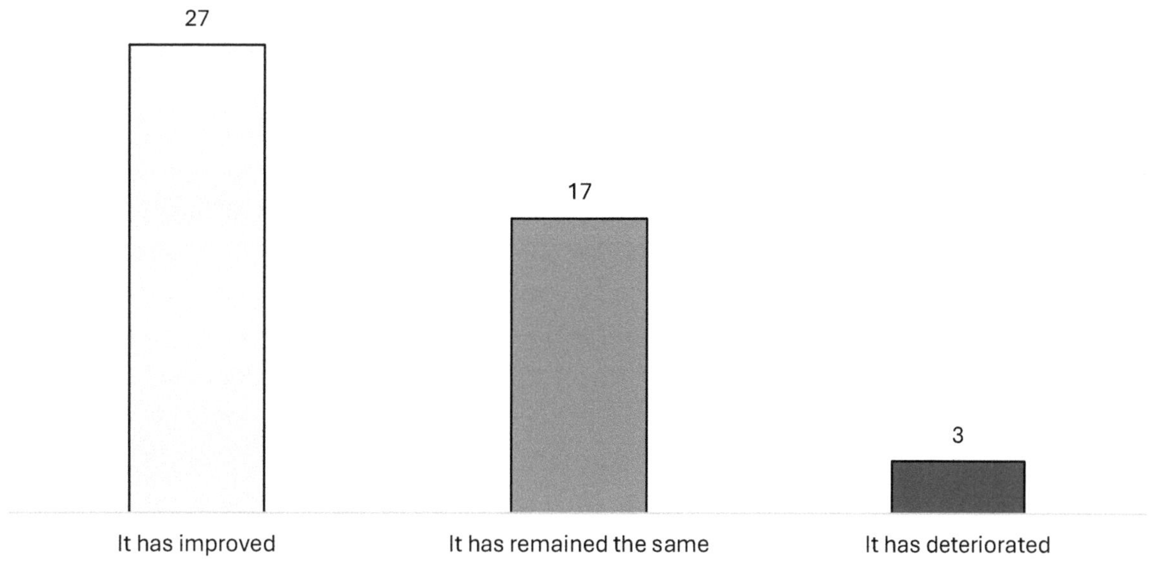

Some examples from countries and regions how the profession of deaf interpreters has improved:

- *Albania*: In the last five years, the profession of sign language interpreters in Albania has developed only modestly. The main change was in 2023, when the Ministry of Justice for the first time recognized interpreters in the National Registry of Translators and Interpreters, certifying four who can work in the justice system. However, for deaf interpreters there are still no regulations, accreditation mechanisms, or state-supported training programs, and most opportunities remain dependent on NGO initiatives such as ANAD. However, in 2025 five deaf interpreters were certified.
- *Austria*: funding has improved.
- *Croatia*: more translation work for deaf interpreters.
- *Denmark*[DI]: Growing visibility and demand. There is increasing recognition of situations where deaf interpreters are essential, such as working with different sign language variations, cross-linguistic communication, or signers with non-standard language. In public settings like hospitals, courts, and social services, the demand for deaf interpreters is also growing. Technology, especially video interpreting, is becoming more common, which creates both opportunities and challenges. At the same time, greater awareness has led to more assignments for deaf interpreters in recent years.
- *Engl. W&NI-ASLI*: The registration for deaf interpreters and translators has grown in the last 5 years. There are more organizations offering training to become a qualified *Intralingual Interpreter*. There is a slow but steady understanding of the importance of deaf interpreters. Due to this growth, there is more research and academic papers being released.
- *Ireland*: With the course for deaf interpreters in 2021 under the Centre of Deaf Studies, the number of deaf interpreters has increased from roughly 5 to over 25. With the ISL

Act 2017, it improved the provision of ISL access, and in some cases, DI/HI Teams are needed. There was a recent research report on Public Services provision, and one of the key recommendations mentioned is to have more ISL Access in-person and to give Deaf clients their autonomy to choose their preferred interpreters.
- *France:* Creation of two master's degrees for translation and one master's degree for intermediation. The COVID pandemic highlighted the work of co-interpreters on TV.
- *Norway*: As some deaf interpreter students have gone through the interpreting programs, they are now more used to having deaf interpreter students and customizing for them.
- *Poland*: The main change within the Association has been a shift in philosophy, from a hearing-centered to a more deaf-hearing collaborative approach. We have become more aware of the importance of signing space, that meetings should not be conducted with hearing people speaking and interpreters translating, but rather that those who know sign language use it directly, with interpretation into Polish provided only when needed. More of our meetings are now conducted entirely in sign language. At the same time, we are aware that our Association does not represent all interpreters in the country. The Polish Association of the Deaf also has an increasing number of vlogs and online materials now interpreted by deaf interpreters rather than hearing interpreters.
- *Serbia*: Over the past five years, the profession for deaf interpreters has seen limited development. The main growth has been in opportunities to work on video translations for online services, while formal recognition, accreditation, and professional pathways remain largely unavailable.
- *Sweden*: The education for deaf interpreters at the folk high school is now permanent and the Stockholm University launched an intramodal interpreter program, bachelor's degree.

The countries that report that the situation has remained the same, give several explanations:

- No deaf interpreters in the country or region (Cyprus, Kosovo, Luxembourg)
- No official recognition or training for deaf interpreters (Greece, Malta)
- No understanding of the role of deaf interpreters (Ukraine[HI])
- No dedicated program or professional development opportunities dedicated to deaf interpreters (Denmark, Lithuania, Slovakia)
- No sign language rights or law (Montenegro)
- Limited funding and training opportunities (Germany)

Three countries report that the situation for deaf interpreters has deteriorated:

- *Estonia*: It does not apply only to deaf interpreters. There is no training program anymore, therefore, the possibilities for both deaf and hearing people to become interpreters is no longer available.
- *Romania*: There are no training and schooling programs for deaf or hard of hearing interpreters.
- *Slovenia*: Deaf interpreters are self-sufficient and do not have any special requirements.

Table 3.1 shows an overview per country/region of the main barriers deaf interpreters face, as well as the changes and opportunities they have.

Table 3.1 Opportunities and barriers for deaf interpreters		
Country / region	**Main barriers**	**Changes / Opportunities**
Albania	Lack of regulation, limited to no funding, limited training	Recognition in justice system registry. New training and certification completed.
Belgium-Flanders	Lack of awareness, education, no full-time working opportunities	Equal payment structures introduced, work with deafblind clients
Belgium-Wallonia	Funding barriers for co-interpreting	Increased demand and accessibility requirements
Bosnia and Herzegovina	Communication barriers; lack of understanding of deaf needs	Increased awareness of accessibility hence opportunities for deaf interpreters
Bulgaria	Limited employment opportunities and funding	Slight improvement in work opportunities
Croatia	Lack of public awareness	Increased opportunities due to international sign interpreting, and demand for translation work
Cyprus	No deaf interpreters exist yet; no training pathways	No opportunities reported
Czech Republic	Limited education and legal recognition	More employment opportunities emerging
Denmark[DI]	Lack of awareness and knowledge among authorities and service requesters about the role and value of deaf interpreters. Lack of self-awareness among hearing sign language interpreters, who sometimes underestimate or overlook the need for deaf interpreters in certain contexts.	Greater recognition and acceptance of deaf interpreters than before. Growing demand in specific contexts, more assignments, and wider awareness of their professional role.
Engl. W&NI-ASLI	Underutilization of Deaf interpreters; funding barriers	Intralingual interpreting opportunities and increase of translation work
Estonia	No training programs	Increased translation opportunities due to accessibility policies
Finland	Institutional restrictions from interpreting service system	Opportunities have decreased
France	Professional role not clearly defined, financial barriers	Increased work in broadcasting and conferences
Georgia	Social stigma	Development of sign language variety
Germany	Financial barriers	Accessibility legislation creating new translation requirements
Germany[DI]	Cost issue and understanding/awareness.	Organization of preparatory courses for the state examination in Darmstadt for deaf people who want to qualify as interpreters. (The quality of preparatory courses is also criticized.)
Hungary	No information reported	Profession officially registered and training started in 2025
Iceland	Lack of training and funding	Few opportunities
Ireland	Limited regular employment opportunities	Increased media and cultural interpreting opportunities
Italy	Cultural barriers and limited recognition, lack of training	Universities program established for deaf and hearing, association membership increase
Latvia	Lack of awareness of need for deaf interpreters	New education opportunity
Lithuania	Lack of education and competencies	No reported change
Luxembourg	No information reported	No opportunities
Malta	Lack of education	No reported change

Table 3.1 Opportunities and barriers for deaf interpreters		
Country / region	**Main barriers**	**Changes / Opportunities**
Montenegro	No deaf interpreters	No reported change
Netherlands	No information reported	No information reported
Norway	Attitudinal barriers and very small professional community	Deaf interpreters recently hired
Poland	Lack of formal education pathways; resistance from hearing interpreters	Accessibility legislation increased translation work
Portugal	Lack of knowledge about the profession; communication barriers	Growing recognition of deaf interpreters' work
Romania	No training programs	Training abroad for professional development
Russia	Low payment	Increased professional promotion and university training
Scotland	Lack of awareness	
Serbia	Lack of training, accreditation and recognition	Opportunities mostly around translation work, irregular opportunities
Slovakia	No official recognition of deaf interpreter role	University training program introduced
Slovenia	Incompetence	No change until legislation improves
Sweden	Limited full-time interpreting opportunities	Growing advocacy and recognition from hearing interpreters
Switzerland-Fr.	Lack of recognition and training	No reported change
Switzerland-It.	Financial barriers	Increased awareness among broadcasters
Turkey	Lack of training, accreditation, institutional awareness	Some NGO-led training and participation initiatives, public institutions and media involvement of deaf interpreters
Ukraine[HI]	Stigma towards deaf interpreters	No information reported

3.3 Hearing interpreters supporting deaf interpreter developments

The respondents were asked how hearing sign language interpreters support deaf interpreters in their professional development and work. Five main themes were identified in the responses (see Table 3.2a and Table 3.2b).

Table 3.2a How hearing interpreters support the work of deaf interpreters	
Theme	**Description**
1. Team collaboration	Deaf and hearing interpreters work together in interpreting teams
2. Training and mentoring	Professional development through workshops and mentoring
3. Organizational support	Interpreter associations supporting deaf interpreters
4. Advocacy	Hearing interpreters promoting the use of deaf interpreters
5. Limited support	Little collaboration due to early stage of the profession

Theme 1: Team Collaboration and Co-interpreting

The most frequently reported form of support involves collaborative teamwork between hearing and deaf interpreters. Respondents described working together on assignments, particularly in complex interpreting contexts where deaf-hearing interpreter teams may improve communication accuracy and cultural mediation. Examples include references to working in tandem, cooperating during assignments, and forming interpreting teams.

Theme 2: Training and Professional Development

Another form of support involves training initiatives and mentoring relationships. In some contexts, experienced hearing interpreters provide informal mentoring or participate in training programs that include deaf interpreters. These initiatives include joint workshops, interpreter education programs involving both deaf and hearing students, mentoring relationships, professional development sessions. In some countries, interpreter training programs are beginning to include deaf and hearing interpreters together, encouraging collaboration from the early stages of professional practice.

Theme 3: Organizational and Association Support

Interpreter associations and professional organizations also play a significant role in supporting deaf interpreters. Several responses describe such associations as including deaf interpreters as members, creating working groups focused on deaf interpreting, supporting networking opportunities, representing deaf interpreters in discussions with governments or institutions.

Theme 4: Advocacy and Promotion of Deaf Interpreters

Some respondents reported that hearing interpreters support deaf interpreters through advocacy and awareness-raising activities. This includes promoting the role of deaf interpreters to clients, recommending deaf interpreters for appropriate assignments, and explaining the benefits of deaf-hearing interpreting teams.

Theme 5: Limited or Absent Support

Despite examples of collaboration and advocacy, several responses indicate that support structures remain limited in some countries. In contexts where deaf interpreters are or the profession is still developing, hearing interpreters may have little experience working with deaf colleagues. Some respondents reported minimal collaboration or limited awareness of the deaf interpreter role, indicating that professional support systems are still emerging.

Table 3.2b How do hearing interpreters support the work of deaf interpreters

Country / region	Team collaboration	Training and mentoring	Organizational support	Advocacy	Limited support
Albania					X
Austria				X	
Belgium-Flanders			X	X	
Belgium-Wallonia	X	X			

Table 3.2b How do hearing interpreters support the work of deaf interpreters					
Country / region	Team collaboration	Training and mentoring	Organizational support	Advocacy	Limited support
Bulgaria	X	X	X	X	
Croatia	X	X			
Cyprus					X
Czech Republic	X	X	X		
Denmark	X				
Engl. W& NI-ASLI		X		X	
Engl. W&NI-NUBSLI					X
Estonia	X	X	X		
Finland					X
France	X				X
Germany	X			X	
Germany[DI]					X
Greece	X			X	
Hungary	X	X	X	X	
Iceland					X
Ireland	X	X			
Italy-ANIMU	X	X			
Italy-ANIOS	X			X	
Kosovo					X
Latvia					X
Lithuania	X				
Luxembourg					X
Malta					X
Moldova	X	X			
Montenegro					X
Netherlands	X		X	X	
Norway	X				X
Poland	X		X	X	
Portugal	X			X	
Romania	X				X
Russia	X				
Scotland		X		X	
Serbia	X			X	
Slovakia	X				
Sweden	X	X			
Switzerland-Fr.					X
Switzerland-Ge.	X				
Switzerland-It.				X	X
Turkey	X				
Ukraine[HI]	X				

3.4 Deaf interpreters in official standards

The profession of sign language interpreters can be mentioned in various official standards, such as qualification frameworks or professional guidelines. This provides a specific status to the profession. The respondents were asked if deaf interpreters are explicitly mentioned or referenced in any official documentation (Table 3.3).

Table 3.3 Are deaf interpreters included in official standards, qualification frameworks, or professional guidelines in the interpreting sector? (n=48)

Yes (n=17)	Partially (n=13)	No (n=17)
Austria	Belgium – Flanders	Albania
Belgium - Wallonia	Bosnia-Herzegovina	Cyprus
Bulgaria	Czech Republic	Denmark[DI]
Croatia	Finland	Iceland
Denmark	France	Italy – ANIMU
England, W&NI (ASLI & NUBSLI)	Georgia	Kosovo
Estonia	Ireland	Luxembourg
Germany	Italy - ANIOS	Malta
Greece	Norway	Montenegro
Hungary	Portugal	Moldova
Latvia	Russia	Poland
Lithuania	Sweden	Romania
Netherlands	Turkey	Serbia
Scotland		Slovakia
Slovenia		Spain
Switzerland – It.		Switzerland – Fr.
		Switzerland – Ge.
		Ukraine[HI]

<table>
<tr><th colspan="5">Table 3.4 Inclusion and exclusion of deaf interpreters in official frameworks</th></tr>
<tr><th colspan="2">Category</th><th>Evidence of Inclusion</th><th>Reasons for lack of inclusion</th><th>Countries</th></tr>
<tr><td rowspan="2">Emerging recognition</td><td>Explicit recognition of deaf interpreters</td><td>Deaf interpreters are explicitly recognized as a professional category</td><td rowspan="2"></td><td>Belgium-Wallonia, Hungary, Latvia, England W&NI-ASLI.</td></tr>
<tr><td>Training-based inclusion</td><td>Deaf interpreters participate in interpreter training programs or qualification processes</td><td>France, Slovenia.</td></tr>
<tr><td rowspan="2">Early inclusive frameworks</td><td>Association inclusion</td><td>Deaf interpreters included in association rules, codes of conduct, or committees</td><td rowspan="2">No official frameworks</td><td>Switzerland-It.</td></tr>
<tr><td>Informal or partial inclusion</td><td>Deaf interpreters included indirectly or through general interpreter standards</td><td>Bulgaria, Belgium-Flanders, Croatia, Estonia, France, Finland, Germany, Lithuania, Norway, Romania, Turkey, Sweden, Turkey.</td></tr>
<tr><td rowspan="2">Structural exclusion</td><td rowspan="2">Absence of formal inclusion</td><td>Based on recommendation to employ deaf interpreters</td><td>No laws or regulations covering deaf interpreting, lack of awareness</td><td>Albania, Bosnia and Herzegovina, England W&NI-NUBSLI, Italy, Luxembourg, Montenegro, Poland, Russia, Serbia, Switzerland-Fr.</td></tr>
<tr><td colspan="2">Absence of deaf interpreters</td><td>Cyprus, Iceland, Luxembourg, Malta, Ukraine[HI].</td></tr>
</table>

The responses reveal three main structural patterns regarding the inclusion of deaf interpreters in official frameworks (Table 3.4).

1. Emerging recognition

In several countries, deaf interpreters are beginning to receive formal recognition through professional registration, association regulations, or the development of dedicated training programs. This stage commonly includes:

- Membership within professional interpreter associations
- The establishment of training programs
- Explicit references to deaf interpreters within professional guidelines

2. Early inclusive frameworks

Some countries incorporate deaf interpreters within existing interpreter standards, regulating them under the same professional frameworks as hearing interpreters. In these contexts, regulations often refer to interpreters in general terms rather than distinguishing between hearing and deaf practitioners. As a result, deaf interpreters may operate under the same ethical standards and professional guidelines as hearing interpreters.

3. Structural exclusion

In many countries, however, deaf interpreters remain outside official regulatory frameworks.

This situation often results from broader structural factors rather than deliberate exclusion. Key barriers include:

- The absence of interpreter training systems
- The lack of national qualification frameworks
- Limited awareness of the deaf interpreter profession
- Hearing-centered models of interpreting practice
- Insufficient funding for deaf-hearing interpreter teams

As a result, deaf interpreters are often employed only through informal recommendations or, in some cases, not employed at all.

Overall, the findings indicate that recognition of deaf interpreters within official standards, qualification frameworks, and professional guidelines remains limited. Further efforts are required to ensure that deaf interpreters are acknowledged and respected as professionals on an equal footing with hearing interpreters.

3.5 Measures to strengthen deaf interpreters' position

The respondents were asked which overall measures they thought would strengthen the position of deaf interpreters in their country or region (Table 3.5). Although responses varied in detail, they broadly clustered around five key areas: recognition, training, structural support, awareness and collaboration.

1. Formal recognition and regulation
A frequently mentioned measure is the need for stronger formal recognition of deaf interpreters within national interpreting systems. Respondents highlighted importance of including deaf interpreters in official standards, qualification frameworks, and professional regulations. In several cases, respondents note that deaf interpreters are not explicitly mentioned in legislation or professional guidelines, which limits their visibility and legitimacy within the profession.

2. Education and training opportunities
Many respondents emphasize the need for dedicated training programs for deaf interpreters. The lack of accessible and specialized education pathways is identified as a barrier to entering and developing the profession. Suggested measures include national training programs, developing deaf interpreter-specific curricula, and ensuring access to continuing professional development.

3. Structural and institutional support
Another recurring theme concerns the need for stronger structural support. Respondents point to the importance of funding mechanisms, employment opportunities, and institutional recognition that would enable deaf interpreters to work more consistently. In particular, the funding of deaf-hearing interpreter teams is identified as an essential component of effective

interpreting practice.

4. Awareness and understanding of the deaf interpreter role
Several responses highlight a lack of awareness among institutions, service providers, and even within the interpreting profession itself about the role and value of deaf interpreters. Increasing awareness of the deaf interpreter profession, its competencies, and the situations in which deaf interpreters are needed is seen as an important step toward improving their professional standing.

Table 3.5 Overall measures needed to strengthen the position of deaf interpreters

Country / region	Key measures
Albania	Regulation, fair pay, training, recognition
Belgium-Flanders	Awareness, dedicated training program
Belgium-Wallonia	Limit access to profession
Bosnia & Herzegovina	Opportunities, inclusion, organization of deaf and hearing interpreters
Bulgaria	Inclusion in public services, awareness, ethical/confidentiality training
Croatia	Inclusion in assignments, advocacy for opportunities
Cyprus	Training programs, good working conditions, cooperation with hearing interpreters
Denmark	National training program
Denmark[DI]	An important need is greater knowledge, as well as more systematic feedback and evaluation from users.
England W&NI (ASLI)	Regulation, education, better commissioning awareness, funding
England W&NI (NUBSLI)	Funding systems (e.g., Access to Work)
Estonia	Training
Finland	N/A
France	Recognize it as distinguished profession
Georgia	Legal changes
Germany	Training, acceptance, funding
Germany[DI]	Universities must be open to deaf students. Raise awareness among board members of interpreter organizations by discussing whether deaf interpreters are requested by hearing colleagues. If not, there is hardly any backing and support for deaf interpreters.
Hungary	Not yet defined
Iceland	Awareness and knowledge dissemination
Ireland	Public awareness, media visibility, funding for training
Italy-ANIOS	Visibility, encourage deaf students into profession
Italy-ANIMU	Equal-level training with hearing interpreters
Latvia	Awareness campaigns, research support
Lithuania	Deaf interpreters interpreting national news, visibility
Luxembourg	Stronger establishment
Malta	Education
Montenegro	Recognition of sign language
Netherlands	N/A
Norway	Increase number and employment of deaf interpreters
Poland	Training/certification, team interpreting, mentoring, awareness, equal access
Portugal	Recognition, inclusion in teams, professional belonging
Romania	Education, bachelor's and master's programs

Table 3.5 Overall measures needed to strengthen the position of deaf interpreters	
Country / region	**Key measures**
Russia	Networking, experience exchange, conferences
Serbia	Training, accreditation, employment pathways, awareness
Slovakia	Legal recognition, acceptance of team interpreting despite funding issues
Slovenia	More training, competencies, structured education
Spain	Official training, collaborative workspaces
Sweden	Government recognition, improved education
Switzerland- Fr.	Official recognition, training programs, collaboration
Switzerland-It.	Recognition of L1M1 interpreters
Turkey	University programs, awareness campaigns, mentorship, community educators
Ukraine[HI]	N/A

3.6 Work and employment of deaf interpreters

The respondents were asked how many deaf interpreters work as interpreters in their country or region. As it is not always possible to give a definite number, each respondent was asked if they could provide an exact number or an estimate. They could also indicate that there are no deaf interpreters in their country or region or that they did not know. The answers are listed in Table 3.6. In addition, Figure 3.3 gives an overview of the exact and estimated number of deaf interpreters.

Table 3.6 Number of deaf interpreters working as interpreters

Country / region	Exact number	Estimated	No deaf interpreters	Don't know
Albania	5			
Austria		5		
Belgium – Flanders		20		
Belgium – Wallonia		20		
Bosnia-Herzegovina			X	
Bulgaria		20		
Croatia		3		
Cyprus			X	
Czech Republic		32		
	16 deaf and 16 hard of hearing.			
Denmark	9			
Engl., W.&NI – ASLI		80		
	ASLI have over 80 deaf interpreter and translator members, but some of these are students.			
Engl., W.&NI – NUBSLI		5		
	Number is based off the last demographic survey in 2017.			
Estonia		1		
	We have only one certified Deaf interpreter but several deaf people who serve in different situations as interpreters, for example interpreting performing arts or translating the Bible.			
Finland		15		
France		9		
	This does not into account translators (less than 10?) and intermediators (around 20 ?).			
Georgia	6			
Germany		35		
Germany[DI]		40		
Greece				X
Hungary			X	
Iceland		2		
Ireland	24			
Italy – ANIMU				X
Italy – ANIOS		10 - 15		
Kosovo				X
Latvia	0			
Lithuania	8			

<table>
<tr><th colspan="5">Table 3.6 Number of deaf interpreters working as interpreters</th></tr>
<tr><th>Country / region</th><th>Exact number</th><th>Estimated</th><th>No deaf interpreters</th><th>Don't know</th></tr>
<tr><td>Luxembourg</td><td>0</td><td></td><td></td><td></td></tr>
<tr><td>Malta</td><td></td><td></td><td>x</td><td></td></tr>
<tr><td>Moldova</td><td>2</td><td></td><td></td><td></td></tr>
<tr><td>Montenegro</td><td></td><td></td><td>x</td><td></td></tr>
<tr><td>Netherlands</td><td>5</td><td></td><td></td><td></td></tr>
<tr><td rowspan="2">Norway</td><td></td><td>5 - 6</td><td></td><td></td></tr>
<tr><td colspan="4">5-6 with a degree. There are others who have taken on assignments, especially in times of need with deaf refugees, but I do not have the numbers for untrained deaf interpreters.</td></tr>
<tr><td rowspan="2">Poland</td><td></td><td>36</td><td></td><td></td></tr>
<tr><td colspan="4">According to the 2023 National Census of Polish Sign Language Interpreters (Spis Powszechny Tłumaczek i Tłumaczy Języka Migowego 2023), Deaf interpreters make up approximately 15% of all interpreters in Poland (36 out of 236 respondents). None of them currently work full time as interpreters, and there are no Deaf interpreters in Poland whose primary or full-time employment is interpreting. Even those who take on many assignments usually have another main job, for example as a PJM lecturer at a university, a trainer running PJM courses within their own company, or a teacher in a school for Deaf students who also interprets occasionally. This corresponds with a general trend from the census showing that 72% of all interpreters, regardless of hearing status, do not consider interpreting their main source of income.</td></tr>
<tr><td>Portugal</td><td></td><td>10</td><td></td><td></td></tr>
<tr><td>Romania</td><td></td><td></td><td></td><td>x</td></tr>
<tr><td>Russia</td><td>40</td><td></td><td></td><td></td></tr>
<tr><td>Scotland</td><td></td><td>4</td><td></td><td></td></tr>
<tr><td>Serbia</td><td></td><td></td><td></td><td>x</td></tr>
<tr><td rowspan="2">Slovakia</td><td></td><td>4 - 7</td><td></td><td></td></tr>
<tr><td colspan="4">It is difficult to give a clear answer to this question, as the first formal education program for interpreters was only established in 2020. Prior to that, interpreting was carried out by a few hard of hearing individuals—an estimate 4 to 5 people—who provided interpreting upon request. Among university graduates from the 2024 and 2025 cohorts, there are only two hard of hearing interpreters so far. In addition, some Deaf individuals began interpreting informally when Deaf Ukrainians arrived in Slovakia because of the war, although they are not officially employed as interpreters.</td></tr>
<tr><td>Slovenia</td><td>6</td><td></td><td></td><td></td></tr>
<tr><td>Spain</td><td></td><td>6</td><td></td><td></td></tr>
<tr><td>Sweden</td><td></td><td></td><td></td><td>x</td></tr>
<tr><td>Switzerland – Fr.</td><td></td><td></td><td></td><td>x</td></tr>
<tr><td rowspan="2">Switzerland – Ge.</td><td></td><td>No estimate number provided</td><td></td><td></td></tr>
<tr><td colspan="4">There are deaf interpreters, but it is not yes established yet like hearing interpreters. There are many deaf translators and many cultural interpreters.</td></tr>
<tr><td>Switzerland – It.</td><td>3</td><td></td><td></td><td></td></tr>
<tr><td rowspan="2">Turkey</td><td></td><td>10</td><td></td><td></td></tr>
<tr><td colspan="4">Although they are not officially registered anywhere, there are approximately 10 deaf sign language interpreters I have seen participating in projects and meetings. However, this number is not exact.</td></tr>
<tr><td>Ukraine[HI]</td><td></td><td>5</td><td></td><td></td></tr>
</table>

Fig. 3.3 Number of deaf interpreters in Europe: 445 (n=34)

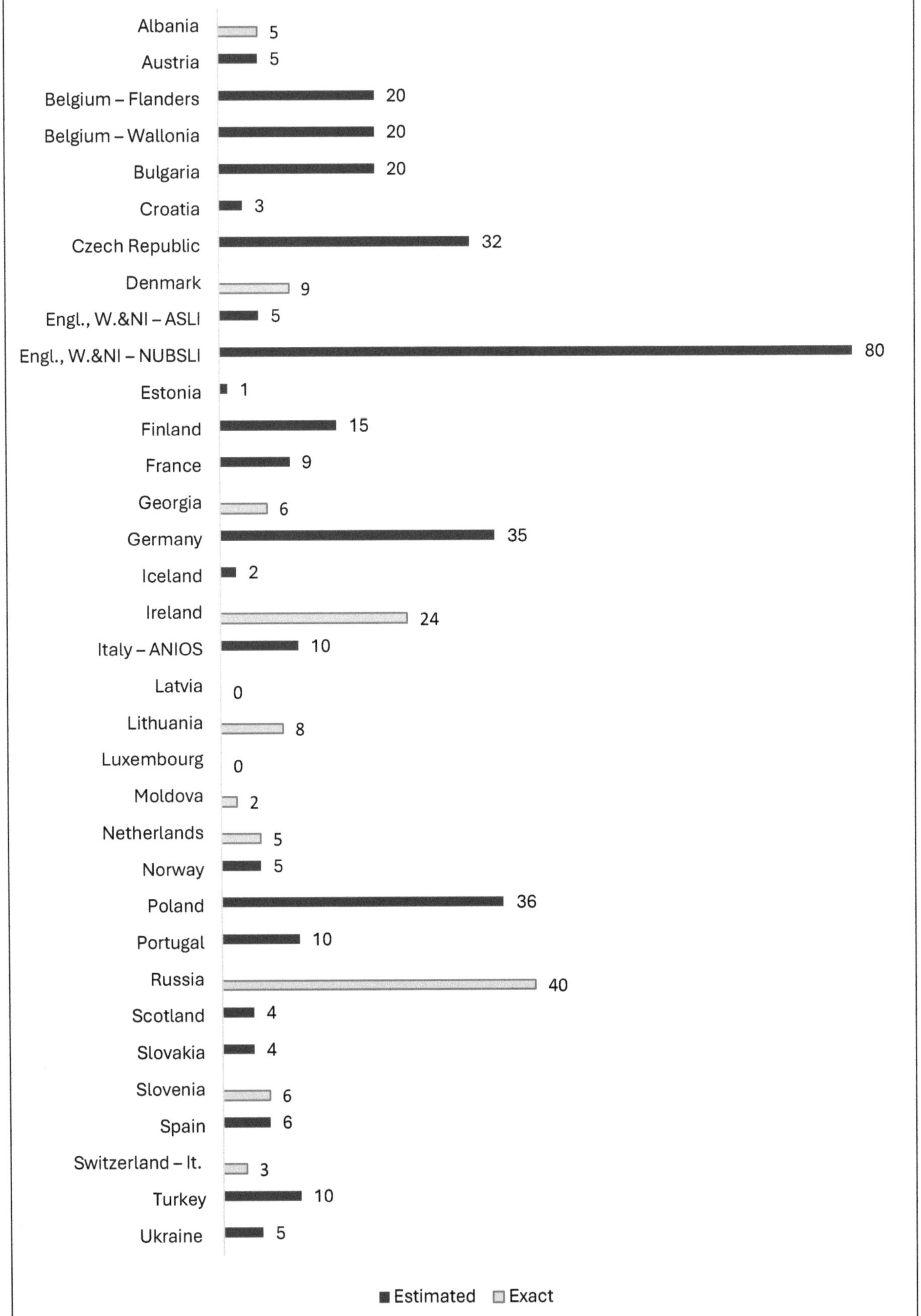

Deaf interpreters are most frequently recruited for international and multilingual settings, as well as for translation services (Table. 3.4). Other settings that are mentioned are:

- Museums and theater (Portugal, Serbia)
- Programs for deaf persons (Switzerland-It.)
- Private sector, for example in advertising (Spain)

Fig. 3.4 Settings deaf interpreters are most recruited for (n=43)

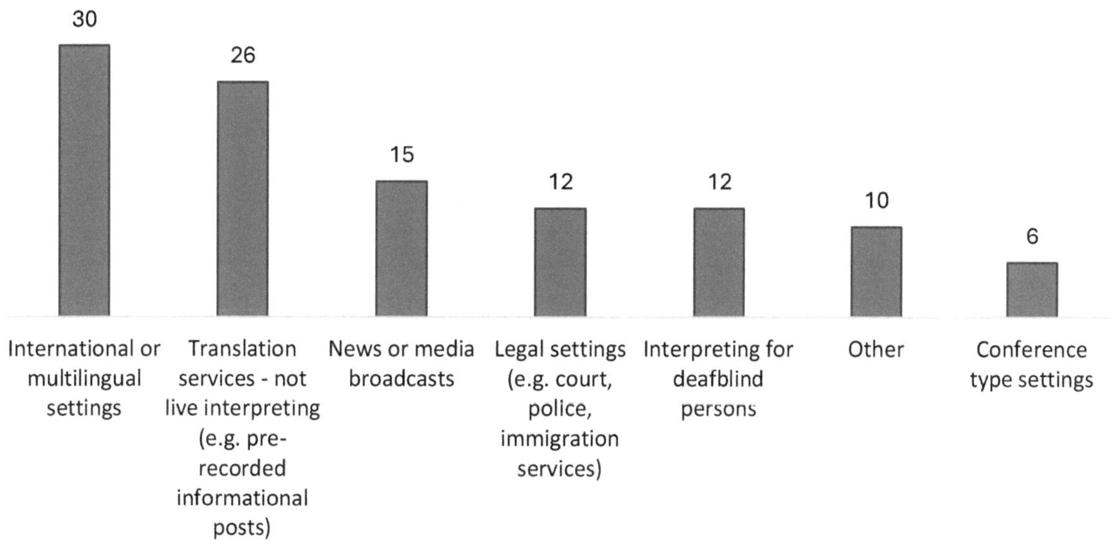

When deaf interpreters work from a spoken to a signed language or vice versa, they work with a hearing interpreter as a co-interpreting team or, for example, from live captions or subtitles (Fig. 3.5). The most common form is as a co-interpreting team. Next to interpreting from live captions, other forms mentioned are interpreting between two signed languages, or as part of a larger sign language interpreting team, or mirror interpreting (deaf interpreter copies the signs of the hearing interpreter or deaf presenter).

Fig. 3.5 How deaf interpreters work (n=42)

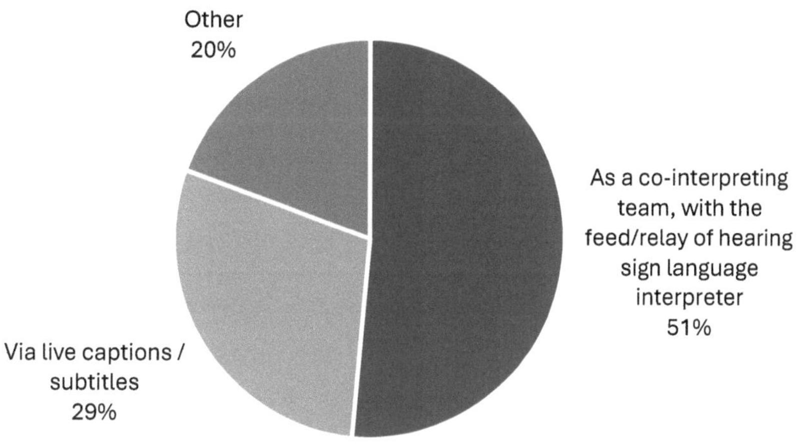

In case of co-interpreting teams there are several options that a deaf interpreter might have:

- Use their personal right to an interpreter to finance the fees of the hearing interpreter
- Decide who they want to work with as co-interpreter

Table 3.7 provides an overview per country if these options apply or not. Only countries and regions that have deaf interpreters and who responded to the question are listed in the Table.

Table 3.7 Options deaf interpreters have when co-interpreting

Country / region	Use their personal right to an interpreter to finance the fees of the hearing interpreter	Decide who they want to work with as co-interpreter	Comment
Albania		x	
Austria	x	x	
Belgium – Flanders	x	x	
Belgium – Wallonia	x		
Bulgaria	x	x	
Croatia		x	
Czech Republic			The pay rate is usually the same for DI and HI; there is no decision on other colleagues' wages. DI can choose their colleague in the same way as HI – the limitation can be capacity and the choice within the working team (the people who have one employer).
Denmark	x	x	
Engl., W.&NI – ASLI	x	x	
Engl., W.&NI – NUBSLI	x	x	
Estonia			The teams are formed as needed.
Finland			Depending on the situation both cases are possible but neither of them apply always.
France		x	
Georgia		x	
Germany / Germany[DI]	x	x	
Greece		x	
Iceland			It's really complicated, mostly used by immigrants.
Ireland			Both interpreters are paid in their irrespective accounts, however in some cases both interpreters can agree how to split the fee e.g. Theatre Interpreting. The option to choose to co-work with a preferred interpreter can be up to the interpreters or booked through an agency.
Italy – ANIMU			Other, not specified.
Italy – ANIOS		x	
Lithuania		x	
Netherlands	x	x	
Norway	x		The deaf interpreters can sometimes decide who they want to work with as a co-interpreter, but not always.
Poland			Usually, it is hearing interpreter who decide who is co-interpreter, or if a team leader is deaf, they decide who will be working in the team.
Portugal		x	

Table 3.7 Options deaf interpreters have when co-interpreting

Country / region	Use their personal right to an interpreter to finance the fees of the hearing interpreter	Decide who they want to work with as co-interpreter	Comment
Russia	x	x	
Scotland		x	Do not know.
Serbia		x	
Slovakia			Co-interpreting team occurred maybe 5 times all in all in our context so far. Not a common practice.
Spain			NA
Sweden		x	Other, not specified.
Switzerland – Ge.			Do not know.
Switzerland – It.		x	
Turkey			The employer is responsible for covering the payment of both interpreters.
Ukraine[HI]			NA

There are three countries that mention a specific payment rule (Fig. 3.6):

- Engl. W&NI - NUBSLI: Based off Access to Work (ATW). If a person has an ATW budget, then they can fund their own interpreters or else, the interpreting service can fund the interpreters. When it comes to private services, there can be private businesses who fund interpreters, wither through ATW or other means.
- Netherlands: The arrangements for deaf NGT interpreters are not quite the same, as there is no formal provision for co-interpreters (although there is for team interpreters). Therefore, they are either considered team interpreters and paid as such, or the deaf interpreter pays for the hearing interpreter's hours based on their own interpreting hours (which they are entitled to as a deaf person). The deaf NGT interpreter then receives the regular rate for the interpreting assignment.
- Russia: There is no legally defined wage level. Payment is made based on a tender and government funding.

Fig. 3.6 Organization of the deaf interpreters' payment (n=40)

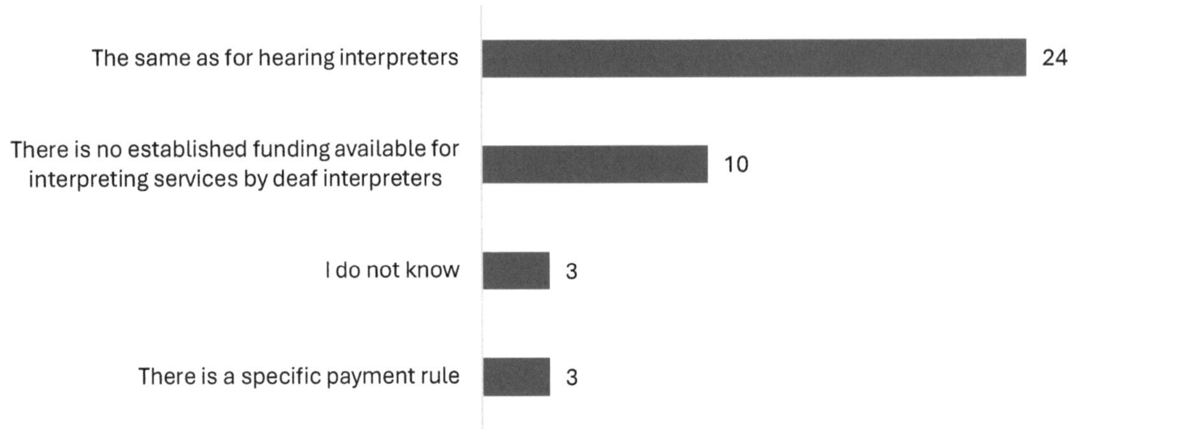

In most countries deaf interpreters work predominantly in translation and some in interpretation (Fig. 3.7).

Fig. 3.7 Distribution of assignments by deaf interpreters (n=41)

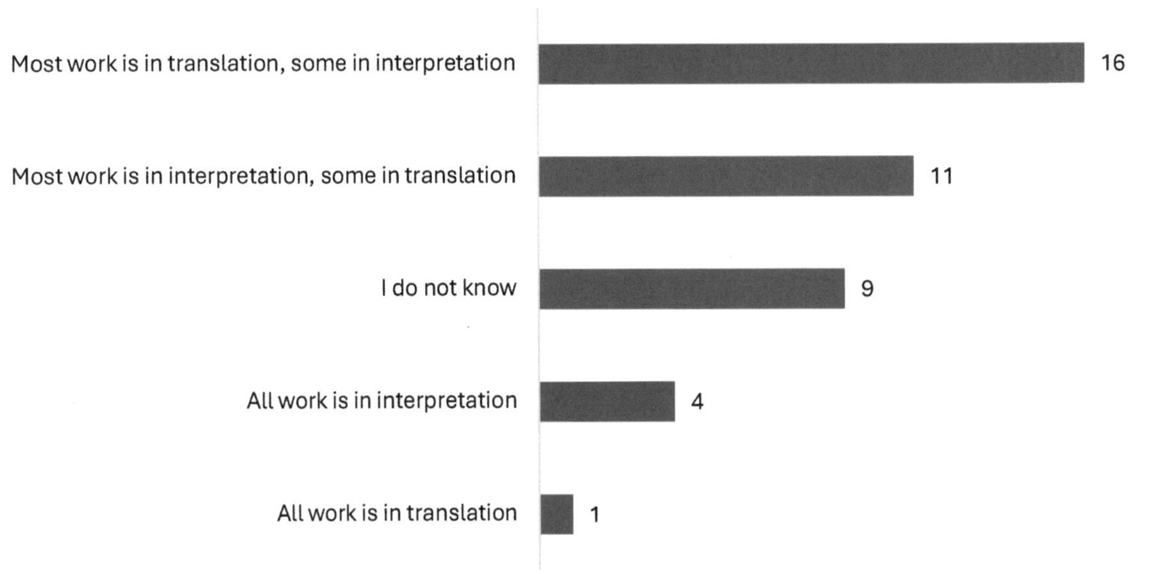

Deaf interpreters face barriers in being recruited as an interpreter (Table 3.8) due to lack of funding, poor awareness of the role of deaf interpreters, absence of training, their inexperience, and a lack of formal recognition. The countries also provided additional comments on the work and employment of deaf interpreters (Table 3.9).

Table 3.8 Main barriers deaf interpreters face in recruitment as an interpreter	
Country / region	**Comment**
Albania	Lack of state regulation; institutions do not provide payment; no continuous professional development
Austria	Need for hearing interpreter in some settings; cost-related factors
Belgium - Flanders	Financial reasons; lack of awareness
Belgium - Wallonia	Limited resources (money); lack of recognition
Bulgaria	Lack of awareness about roles and qualities
Croatia	Lack of awareness; misconceptions about no needing qualification
Czech Republic	Majority of work limited to community interpreting
Denmark[DI]	Lack of awareness among authorities, service requesters, and hearing interpreters.
Engl.W&NI - ASLI	Financial concerns; lack of awareness; limited availability
Engl.W&NI - NUBSLI	High qualification costs; restrictive government systems
Estonia	Not enough assignments for full-time work
Finland	Restrictions by KELA and legislation
France	Lack of recognition; absence of training; limited financial support
Georgia	Stigma
Germany	Financial reasons; gatekeeping by hearing colleagues
Greece	Lack of training; lack of understanding; cost issues
Iceland	Funding; education; lack of understanding
Ireland	No current training courses; limited accreditation pathways
Italy - ANIMU	Inadequate training; funding issues; preference for single interpreter
Kosovo	No data
Lithuania	Lack of education
Luxembourg	No local deaf interpreters; multilingual complexity
Moldova	Limited experience; no interpreting center; very few interpreters
Netherlands	No data
Norway	Under communication of need; limited assignments; traditional practices
Poland	Systemic barriers; limited assignments; lack of advocacy; limited recognition
Portugal	Communication barriers; inexperience; difficulty negotiating conditions
Romania	Lack of qualifications; lack of collaboration; prejudice
Russia	No legal qualification standards
Scotland	Lack of awareness and understanding
Serbia	No accreditation; legal restrictions; limited opportunities; no professional development
Slovakia	No tailored training; lack of legal recognition
Slovenia	Users do not see need; cost of two interpreters
Spain	Lack of awareness; prejudice; limited job opportunities
Sweden	No recognition; limited income opportunities; reliance on other jobs
Switzerland – Fr.	No data
Switzerland – Ge.	No official studies; limited training
Switzerland – It.	Cost; need for double interpreters
Turkey	Communication difficulties; perceived high cost
Ukraine[HI]	No training programs; lack of understanding of role

Table 3.9 Overall comments on the work of deaf interpreters	
Country / region	**Comments**
Austria	In terms of payment rule, it depends on the field; in Austria we generally face the challenge of different stakeholders managing the budgets etc.
Croatia	In Croatia, Deaf interpreters primarily work with deafblind individuals, at conferences, and in translation.
Czech Republic	We don't have any data on the assignments distributed for deaf persons working as interpreters and/or translator. We would say it can be about half and half.
Engl. W&NI - ASLI	We do not know officially the proportion of work that is interpreting or translation and have answered this based on views provided to me by deaf interpreters and translators.
Estonia	Some deaf interpreters only work in translations, some do both, depends on the interpreter and the assignments.
Finland	90 % of the interpreting services for individuals is provided by KELA, the Social Insurance Institution of Finland, and interpreting assignments distributed by KELA's Centre for Interpreting Services. However very few assignments are distributed to deaf SLIs. If any assignments are given to deaf SLIs, both translation and interpretation assignments are possible, but I don't have the ratio. The remaining 10 % of interpreting services needed in Finland is ordered directly from interpreting service providers who can assign SLI's according to the need. Both translation and interpretation work is assigned for deaf SLI's, but I don't have information on the ratio between them.
France	We are at a turning point in the profession where Deaf interpreters, translators and intermediators need to agree on their respective field of expertise and intervention. Moreover, to support the development of Deaf interpreters, it is crucial to work closely with universities to create specific trainings.
Iceland	It's new, not well known in Iceland, lack of education/training, lack of understanding from the hearing community.
Ireland	Most deaf Interpreters work mainly in translation services such as Public Broadcasting e.g. News channel, Public Videos and organizations. Several deaf Interpreters are involved in the Arts, such as Concert interpreting along with a co-interpreter, and teleprompter. Co-interpreting is used in Legal, Medical and Conference situations. Most deaf interpreters have a main job outside of interpreting, as it is not a regular paid job, so this affects their availability, and limits their schedules.
Luxembourg	Payment of DI is done according to German rules and regulations. We basically follow the German system. However, everyone is entitled to ask for the conditions and payment they would like. I will check with the ministry how this has been done (translation / interpretation).
Moldova	We do not have deaf interpreters; we are going to develop this service.
Poland	STPJM has created a dedicated section for Deaf interpreters to post announcements and invite hearing colleagues to join projects, which helps promote Deaf-led collaboration. Yet such situations remain rare compared to the prevailing practice where hearing interpreters are the ones initiating or controlling assignments. This imbalance continues to limit the visibility and professional growth of Deaf interpreters in Poland.
Portugal	It is a visual and cultural asset. It enriches the work of the interpretation teams.
Romania	In Romania, very few people know that there are deaf interpreters. There are no schooling options for deaf interpreters. I can sadly say that for the Romanian state, this category of interpreters does not even exist.
Russia	Deaf interpreters usually work in pairs with a hearing interpreter.
Serbia	In video translations, it is common for a hearing interpreter to first translate the text into sign language, and then a deaf interpreter adapts that translation into a more natural, native form of sign language.
Slovakia	Approximately 4 to 5 deaf interpreters are employed in the regional government (Slovakia has 8 regions / districts) providing interpreting services for clients from that region, but others work upon request not having this as a full-time work.
Switzerland – Fr.	There are two distinct categories within our context: deaf translators, who translate from written French into French Sign Language (LSF), and deaf interpreters, who interpret from foreign sign languages into French-Swiss Sign Language (LSF-CH).
Turkey	The concept of a deaf sign language interpreter is still very new in our country; therefore, there are no established standards regarding payment, working conditions, or similar matters. Although they generally share similar conditions with hearing interpreters, they still have very limited employment opportunities.

Table 3.9 Overall comments on the work of deaf interpreters	
Country / region	**Comments**
Ukraine[HI]	Most of the deaf interpreters' work is done on a voluntary basis if it is carried out at events organized by the National Society of the Deaf or Deaf Sport Federation. There are known cases of hard-of-hearing people working as interpreters between spoken and sign languages at formal events (seminars, trainings, conferences, forums, and staff meetings).

3.7 Training and education of deaf interpreters

How deaf interpreters are trained varies greatly across national contexts (Fig. 3.8; Table 3.10). Overall, the responses show that training systems for deaf interpreters remain uneven and underdeveloped in many countries. The most significant challenges include the lack of formal programs, limited accessibility of existing education pathways, insufficient specialized curricula, and restricted opportunities for professional development.

Although many countries report limited or absent training opportunities for deaf interpreters (Table 3.11), there are several examples of emerging good practices. These include university-based training programs, mentoring initiatives, pilot projects, and continuing professional development opportunities.

Fig. 3.8 How deaf interpreters are trained (n=47)

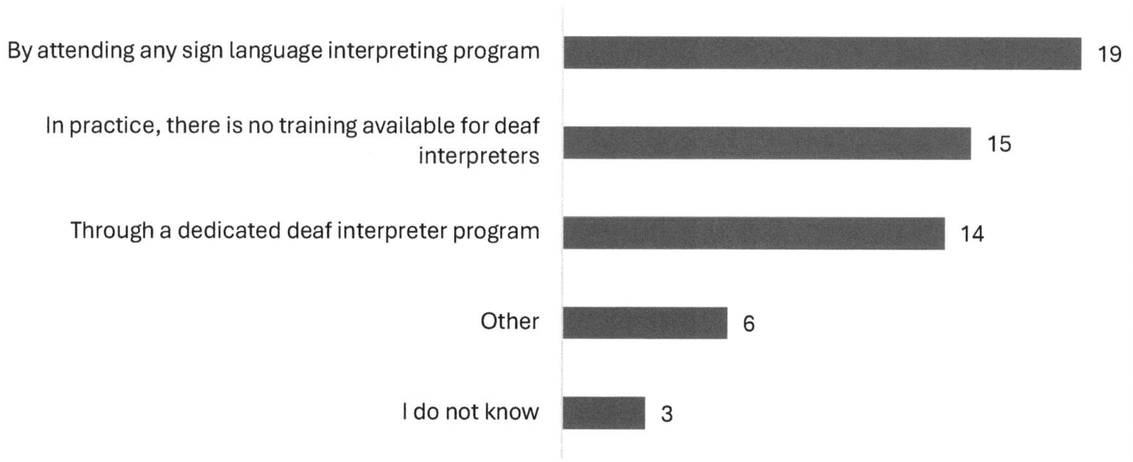

In most countries and regions interpreters' education is designed primary for hearing interpreters and deaf students are expected to meet the same requirements despite differences in linguistic access and educational background (Fig. 3.8). The absence in a curriculum of modules specifically tailored to the role of deaf interpreters does not sufficiently address the specific competencies required for deaf interpreting, such as working in deaf-hearing interpreter teams or working from live text (e.g. captions, subtitles).

In practice, training opportunities are either non-existent or only offered sporadically through short courses, projects, or NGO initiatives rather than through established university programs or nationally recognized systems (Table 3.10).

Even where training programs exist, they may not be fully accessible to deaf students. Several responses noted that university level requirements, language expectations, or admission criteria can make interpreter education difficult to access for deaf applicants. In addition, deaf interpreters may have fewer opportunities for internships, mentoring, or practical experience during training.

Respondents also mention the shortage of qualified deaf trainers, pedagogical resources, and specialized teaching materials, which further limits the development of training programs.

Table 3.10 Overview of available training deaf interpreters can enroll in

Country /region	Type of training G-SLI= General SLI training DDI=Dedicated deaf interpreter training OTH=Other	
Albania	G-SLI	Occasional courses provided on demand.
Austria	DDI	
Belgium – Flanders	DDI	Until now, there has only been one dedicated deaf interpreter program, which is now finished. We do not know if there will be other training opportunities in the (near) future.
Belgium – Wallonia	DDI	
Bulgaria	G-SLI	
Czech Republic	G-SLI	University interpreter education exists, although most deaf interpreters gain training through organizations where they work.
Denmark	DDI	
Engl. W&NI – ASLI & NUBSLI	G-SLI, DDI	
Estonia	G-SLI	Past shared training for deaf and hearing interpreters training.
Finland	G-SLI	
France	G-SLI OTH: Private course	France does have a training program (master's degree) for translators, and another one for intermediators. But there is not specific training for Deaf interpreters. In most cases, translators are the one to consider themselves best suited for co-interpretation. For this setting, there is a 2.5-day program dispensed by one French Deaf interpreter who got training abroad and one French Deaf translator/interpreter who got trained by this first person and got experience over time.
Georgia	G-SLI	
Germany	DDI	
Hungary	DDI	
Iceland	G-SLI	University level.
Italy – ANIOS	G-SLI, DDI	
Latvia	OTH: Pilot training	Pilot college program with one deaf student.

Table 3.10 Overview of available training deaf interpreters can enroll in

Country /region	Type of training G-SLI= General SLI training DDI=Dedicated deaf interpreter training OTH=Other	
Lithuania	G-SLI	
Luxembourg	G-SLI	Deaf professionals trained abroad (University of Cologne).
Malta		Occasional courses provided on demand.
Netherlands	G-SLI	Five deaf students graduated from the general SLI training program, in which the curriculum was adapted to their profile.
Norway	G-SLI	Training programs exist but are mainly designed for hearing interpreters, with deaf students participating in them.
Poland	OTH: mentoring	Mentoring program through the national interpreter association.
Portugal	DDI	Course.
Russia	DDI	Newly introduced national training program by the government.
Scotland	G-SLI	
Slovakia	G-SLI	There is a training program at Trnava University that is open to both hearing and deaf students; however, it follows a single curriculum without a dedicated component specifically designed for deaf interpreters.
Slovenia	G-SLI	No dedicated program but interpreter training exists.
Sweden	DDI	Formal education program for deaf interpreters exists, although it is relatively short and separate from hearing interpreter training.
Switzerland- It	DDI	At UNIGE there was a deaf translators/interpreters dedicated program but now is no more available for LIS users.
Turkey	G-SLI	No dedicated program but interpreter university training exists from which deaf students have graduated.
Ukraine[HI]	G-SLI	

Switzerland French region states:

> "There is currently no interpreter training available, even for hearing interpreters, although the demand is clear. Developing such programs is resource-intensive, and regional priorities tend to focus on more immediate needs. Consequently, interpreters are often trained abroad or recruited from neighboring countries such as France. The role of deaf interpreters is still emerging and not yet recognized in official frameworks. There are no established qualification pathways or standards defining their role."

The University of Geneva in the Switzerland-French region is aiming to secure funding for a new program (see 4.3).

Table 3.11 Countries reporting no or very limited training for deaf interpreters

Country / region	Comment reported
Albania	No national interpreter program.
Belgium-Flanders	No dedicated deaf interpreter program now.
Croatia	No formal education. Deaf interpreters in the interpreter association can attend all the education the association offers on equal footing with their hearing colleagues.
Cyprus	No training available.
Czech Republic	No specific training for deaf interpreters.
Denmark	No continuous official national training program for deaf interpreters.
Denmark[DI]	Some hearing sign language interpreters do not acknowledge that the training for deaf interpreters is nearly equivalent in structure and content to their training.
Estonia	Past shared training for deaf and hearing interpreters training.
Greece	No training available.
Ireland	A one off one year training program was run in Trinity College Dublin in 2021/22. Currently available are CPD courses.
Italy - ANIMU	In practice, no training dedicated for deaf interpreters.
Latvia	Only one pilot. No training dedicated for deaf interpreters.
Luxembourg	No national training available for deaf interpreters, trained abroad.
Malta	No training dedicated for deaf interpreters.
Moldova	No training dedicated for deaf interpreters.
Montenegro	No training dedicated for deaf interpreters.
Poland	No training dedicated for deaf interpreters.
Romania	No training dedicated for deaf interpreters.
Serbia	No training dedicated for deaf interpreters.
Slovakia	No training dedicated for deaf interpreters.
Slovenia	No training dedicated for deaf interpreters.
Spain	No training dedicated for deaf interpreters.
Switzerland-Fr.	No training dedicated for deaf interpreters.
Switzerland-Ge.	No training dedicated for deaf interpreters.

Overall, there are relatively very few trained deaf interpreters (Table 3.12). Those that are trained are mainly trained to interpret as a co-interpreting team, together with a hearing sign language interpreter as a feed/relay interpreter (Fig. 3.9).

Table 3.12 Number of trained deaf interpreters (n=34)

Country /region	Exact	Estimate	I do not know
Albania	5		
Austria		4	
Belgium – Flanders		20	
Belgium – Wallonia		20	
Bosnia-Herzegovina			x
Bulgaria		20*	
Cyprus			x
Czech Republic			x
Denmark[DI]	12		
Engl. W&NI – ASLI		60	
Engl. W&NI – NUBSLI		5	
Estonia	2		
Finland		10	
France		9	
Georgia	6		
Germany / Germany[DI]		35-40	
Hungary		1**	
Ireland	26		
Iceland	1		
Italy – ANIOS		10-15	
Kosovo			x
Latvia	0		
Lithuania	2		
Montenegro			x
Netherlands	5		
Norway		5-6	
Portugal	7		
Russia		500	
Scotland			x
Sweden		40	
Switzerland-Fr.			x
Switzerland-Ge.			x
Switzerland-It.	5		
Turkey	5		

**Trained but not in the national register*

***Trained abroad and unable to register in the national registry*

Fig. 3.9 Are deaf interpreters trained to interpret via live captions and/or as co-interpreter with a hearing interpreter?

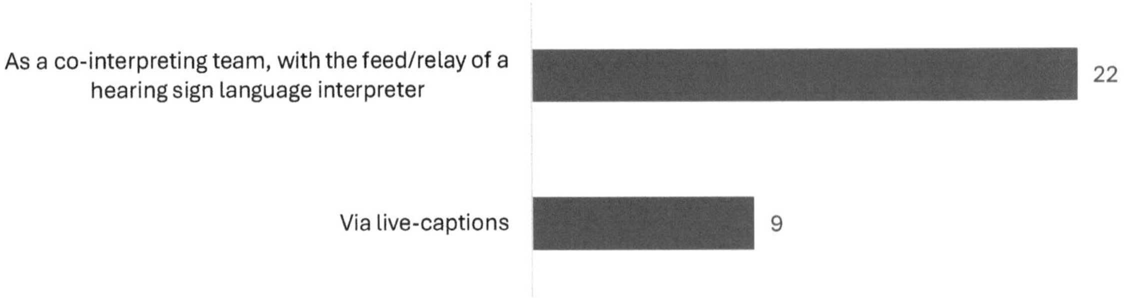

Beyond initial training, respondents also noted the lack of continuing professional development opportunities for deaf interpreters (Table 3.13). In some countries, training sessions are offered only occasionally or irregularly, making it difficult for practitioners to maintain and develop their professional skills. Where continuing professional development opportunities exist, they are often shared with hearing interpreters and may not address the specific professional needs of deaf interpreters.

Despite the challenges identified, some responses highlight early efforts to develop training pathways. These include pilot programs, mentoring initiatives, and university training opportunities that include deaf students. In some countries, organizations or associations have begun to develop new programs aimed at strengthening professional training and supporting collaboration between deaf and hearing interpreters.

Poland, for example, mentions the following in their development of training for deaf interpreters:

> Training opportunities for Deaf interpreters in Poland are still developing and remain limited in number and scope. Within the Association of Polish Sign Language Interpreters (STPJM), training groups are usually mixed, reflecting the overall membership composition—typically no more than 20% of participants are Deaf. While this model promotes shared learning, it also means that deaf interpreters rarely have access to courses designed specifically for their needs or delivered fully in PJM. Among deaf interpreters, there is also a common perception that they are invited to assignments mainly when the task is particularly difficult or when hearing interpreters are unable to manage it, which reinforces unequal professional dynamics. A positive change in 2025 was the implementation of a new STPJM mentoring training program for PJM interpreters, with half of the participants being Deaf and half hearing. This initiative will make it possible to offer mentoring—both within and outside the Association—conducted by qualified Deaf and hearing mentors, marking an important step toward balanced professional development.

Table 3.13 Rating by deaf interpreters

How would a deaf interpreter classify the available interpreter training for deaf interpreters? (n=33)

1 Very bad	2 Bad	3 Neutral	4 Good	5 Excellent
20	4	5	2	2
61%	12%	15%	6%	6%
Albania Belgium-Flanders Bosnia-Herzegovina Estonia France Greece Italy-ANIMU Latvia Lithuania Luxembourg Malta Moldova Poland Romania Serbia Slovakia Slovenia Spain Switzerland-Ge. Ukraine[HI]	Bulgaria Germany Germany[DI] Engl. W&NI-NUBSLI	Belgium-Wallonia Croatia Ireland Norway Sweden	Denmark[DI] Iceland	Russia Switzerland-It.

How is the access to mentoring, and professional development opportunities for deaf interpreters? (n=33)

1 Very bad	2 Bad	3 Neutral	4 Good	5 Excellent
22	6	3	0	2
67%	18%	9%	0%	6%
Albania Belgium-Flanders Belgium-Wallonia Estonia Germany Greece Iceland Latvia Luxembourg Malta Moldova Poland Romania Serbia Slovakia Slovenia Spain Sweden Switzerland-Ge. Switzerland-It. Ukraine[HI]	Bosnia-Herzegovina Croatia Germany[DI] Italy-ANIMU Lithuania Norway	Engl. W&NI-NUBSLI France Ireland		Denmark[DI] Russia

3.8 Current main barriers and future steps

Since the 2020 publication the situation for deaf interpreters in Europe has improved slightly. Deaf interpreters as a profession have gained more attention and employment and training opportunities have increased.

However, deaf interpreters continue to face major barriers and have very few and limited work and educational opportunities compared to hearing interpreters. The barriers can be clustered into three overarching categories:

Systemic Barriers

- Legal recognition
- Training systems
- Institutional frameworks

Economic Barriers

- Cost perceptions
- Funding limitations
- Market demand

Socio-cultural Barriers

- Awareness
- Attitudes and stigma
- Professional power dynamics

The challenges faced by deaf interpreters are not individual but systemic. The profession is constrained by weak institutional foundations, misunderstanding of its value and economic structures that favor hearing interpreters. To address these barriers countries need to focus on awareness raising, policy reform, investment in training and the structural inclusion of deaf interpreters in interpreting systems.

Part 4 – Training & Education

4.1 Educational Programs

Training and education of sign language interpreters is an important of the professionalization process. In this edition new questions were added on the content of the educational programs, such as prerequisites to apply to a program, the application of CEFR,[21] internships, number of students enrolled, and curriculum adaptations for deaf students. Not all respondents were able to answer to these detailed questions, either due to for example the high number of programs (e.g. Germany, Italy), or simply because they did not have access to this detailed information (e.g. England, W&NI, Netherlands).

However, the overall reported data gives insight into the developments over time as well as new trends. During the past decades educational programs for sign language interpreters have developed from short training programs of a few months to curriculums at higher education institutions that are typically three years or more. There are currently more than eighty training possibilities for sign language interpreters in Europe (Table 4.1), up from 2020 when there were around seventy programs. The programs vary per country and language region and from short courses to university master's degree programs.

Most current formal interpreting programs have a duration of one to four years depending on their academic level. Compared to the 2016 and 2020 survey, several programs increased their duration and raised the level of their degrees (Table 4.1). The number of master and bachelor level programs increased and those at lower level at a folk high school or certificate format decreased. The 2020 survey made no distinction between master and bachelor level programs at university or at the university of applied sciences. The current survey does make this distinction now (Fig. 4.1). Altogether the number of programs at master level increased to nineteen programs (sixteen at university and three at university of applied sciences) and at bachelor level to forty-one programs (twenty-one at university and twenty at university of applied sciences).

Table 4.3 provides a detailed overview of all the sign language interpreting programs across Europe. The website links to each program are listed on www.mayadewit.nl/european-study.

---

[21] https://www.coe.int/en/web/common-european-framework-reference-languages/level-descriptions

Table 4.1 Number of interpreter training possibilities: 82									
Country / region	MA at uni.	MA at uni. of applied sciences	BA at uni.	BA at uni. of applied sciences	post-graduate	Vocational / folk high school	certificate/ course	other	TOTAL
Austria	1		1	2				2	6
Belgium-Flanders					1				1
Belgium-Wallonia	1								1
Bulgaria							1		1
Czech Republic			1			1			2
Denmark			1						1
England, Wales & NI	1		1		1				3
Finland	1	2		2				1	6
France	5								5
Georgia							1		1
Germany	2	1	2	5					10
Greece							1		1
Hungary							2		2
Iceland			1						1
Ireland				1					1
Italy	1		5				5		11
Latvia								1	1
Lithuania				1					1
Netherlands				1					1
Norway			3						3
Poland	1								1
Portugal				3					3
Romania					1				1
Russia			1	2					3
Scotland	2								2
Slovakia			1						1
Spain			2						2
Sweden			1			2	1		4
Switz.-Ge				1					1
Switz.-It.			1						1
Turkey	1								1
Ukraine[HI]				2			1		3
TOTAL	**16**	**3**	**21**	**20**	**3**	**3**	**12**	**5**	**82**

Fig. 4.1 Number of educational programs in Europe per educational level (n=82)

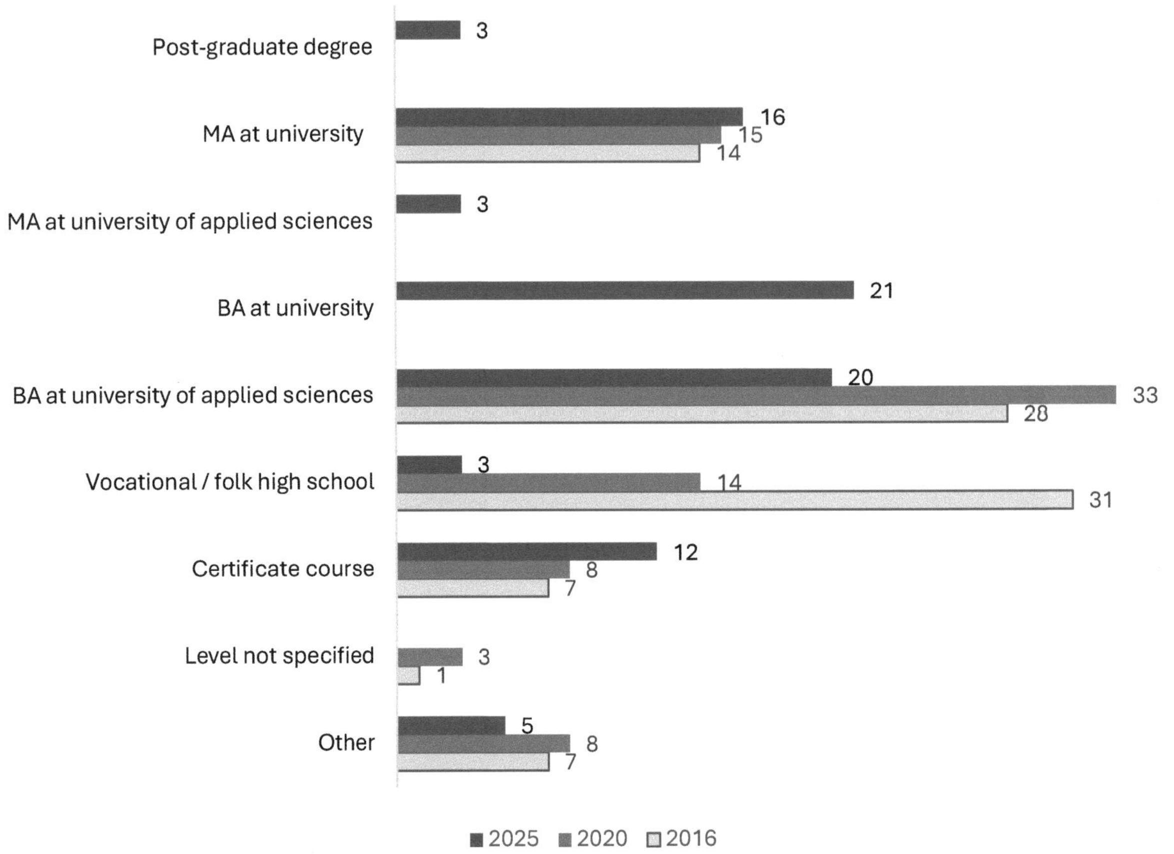

Note: In the 2020 edition no differentiation was made between a BA and MA at university or at a university of applied sciences. MA and BA Postgraduate trainings were not specified in the 2016 edition

4.2 Countries and regions without a program

Not only was there an increase in the number of educational programs, but countries also reported the closing of programs. For example, in Belgium-Flanders there where previously three sign language interpreter programs at vocational and bachelor level, which are now closed. The closure of the programs was due to the decision by the body on professional qualifications in Flanders, which determined that the profession of VGT interpreters was to be at level 7 (academic master). This was approved by the Flemish government on 18 September 2020.[22] As a result of this decision the training of VGT interpreters changed significantly.

[22]https://app.akov.be/pls/pakov/f?p=VLAAMSE_KWALIFICATIESTRUCTUUR:BEROEPSKWALIFICATIE::::1020:P1020_BK_DOSSIER_ID:2102

Currently Belgium-Flanders only has a post-graduate degree program at the university of KU Leuven (Antwerp Campus).[23] In Belgium there was also a dedicated deaf interpreting program of 1.5 years organized by the same university, which ran from January 2022 to September 2023. This was a temporary program and had eighteen graduates.[24]

The program in Estonia was closed some years ago. The national interpreter organization is engaging in discussions with the state and the universities to start a new training. In the meantime, the interpreter association organizes different training projects for interpreting in different areas, such as interpreting in performing arts or relay interpreting, but these are mainly one-time projects.

Table 4.2 gives an overview of all the countries that do not have a formal interpreter training program. Nearly all countries are in the eastern part of Europe. Exceptions are Malta and Luxembourg, the smallest EU member states by population. None of the Balkan countries have an established, formal education for sign language interpreters. If there is any training, these are typically short courses, lasting less than two months. In Romania they just started a weekend course at Bucharest University.

[23] https://onderwijsaanbod.kuleuven.be/opleidingen/n/SC_53777207?faseIds=1
[24] https://www.youtube.com/watch?v=MYByEY3fsjE

Table 4.2 Countries that do not have formal interpreter training programs	
Country / region	**Comments reported**
Albania	Training is primarily provided by the Albanian National Association of the Deaf, which offers a program accredited by the National Agency for Education and of the Deaf (ANAD), which offers a program accredited by the National Agency for Education and Vocational Training. The last training was from 2020 to 2024. Certification took place in 2025 as part of a project by the Finnish Association of the Deaf.
Bosnia-Herzegovina	No information.
Bulgaria	No training for sign language interpreters. There is a course in sign language.
Croatia	The national association organizes occasional training sessions and sign language interpreter workshops.
Cyprus	Most interpreters receive their training in Greece.
Estonia	The former interpreter training program was discontinued several years ago.
Kosovo	Interpreter training has been organized by the national deaf organization.
Luxembourg	Interpreter training is linked to the regional framework of Rheinland-Pfalz through BGSD. Only continuing education opportunities are available within this framework.
Malta	Interpreter training courses are supported through ad hoc government funding.
Moldova	Interpreters are typically individuals who are children of deaf adults (CODAs) or individuals who have learned sign language independently and subsequently developed their skills through involvement with the Deaf Association.
Montenegro	Interpreters generally acquire their skills through interaction with deaf persons.
Romania	No full degree program for sign language interpreters. In October 2025 a weekend course started at the University of Bucharest, which will finalize in June 2026.
Serbia	Interpreters are primarily trained informally through community-based experience and mentorship from more experienced interpreters.
Switzerland-Fr.	Local interpreters receive interpreter training in France.

4.3 Past and new programs

In September 2020, a new educational program for sign language interpreters was established in Slovakia. Trnava University offers a bachelor's program at their Faculty of Education, which focuses on language and interpreting skills, as well as Deaf Studies.

In September 2021, the Faculty of Translation and Interpreting (FTI) at the University of Geneva started a pilot project financed by the PROCOM foundation and the Swiss Federation of the Deaf (SGB-FSS):

> This was a three-year bachelor program on multilingual communication (180 ECTS credits) where students worked on their sign language skills, deaf culture, and so on. Students who already had a diploma in translation, interpreting or another diploma that the university considered to be equivalent could attend only the LSF or LIS courses and get a Certificate of communication of 60 ECTS credits. The first cohort started in 2021 with LSF, the second one started in 2023 with LIS.

> The idea was then to prepare a two-year follow-up that would specifically train interpreters. We now have had external fundings to set up a continuous education program for hearing interpreters. It is a two-year program, and it is a Master of Advanced Studies (MAS - 60 ECTS credits). Hopefully, it will start in January 2027.
>
> Concerning our deaf students, we set up a continuous education training that lasts 2 years and is a Diploma of Advanced Studies (DAS, 30 ECTS credits) who trains them in general translation (year 1) and audiovisual translation (year 2). We are in the process of thinking of setting up a training for deaf interpreters that could have some joint classes with our hearing cohort.[25]

As of the winter semester 25/26 the BA and MA degrees at the University of Graz with sign language translation are also open for deaf students to attend. The curriculum changed and funding for interpretation is secured for all university settings in Graz.

In Turkey they are considering opening a new program, these are their considerations:

> To open a new program in Türkiye, there must be enough teaching staff with appropriate academic titles in the relevant field. Therefore, before establishing a new department, a master's program is first opened to train qualified teaching staff in that specific field. Through this process, individuals first receive a master's education and later obtain a doctoral degree. The establishment of the master's Program in Sign Language Interpreting was also carried out in accordance with these requirements. The department was established and began admitting students in the 2019-2020 academic year; however, it has not admitted any students for the past two years.
>
> About the Program: The Master's Program in Turkish Sign Language Interpreting has been designed to establish the academic foundation necessary for doctoral-level studies that approach Sign Languages and Turkish Sign Language Interpreting as an independent and scientific field of research, like examples in universities around the world.

Table 4.3 gives an overview of all training formally established training programs. Training and educational opportunities for deaf interpreters are further detailed in (as well as in chapter 3.7).

[25] Information received by email from Irene Strasly, project coordinator, 21 April 2026

Table 4.3 Overview of sign language interpreter training programs in Europe

Country / region	Program	Institute, City	Years and FT/PT	ECTS	# students enrolled	Status[26]	Public/ private
Austria	Bachelor and Master Translation Studies - Sign Language[27]	Department of Translation Studies at the University of Graz (ITAT), Graz	FT/PT BA: 3 yrs MA: 2 yrs	BA: 360 MA: 120	BA: 30-40 MA: 15-20	a,b,c	public
	Fachausbildung Gebärdensprachdolmetschen (GESDO)	GESDO (funded by a budget from Upper Austria, County), Linz	FT 3 yrs	180	12-18	a,b,c	private
	MODUS (Dolmetschen und Übersetzen für Österreichische Gebärdensprache, Deutsch und International Sign)	University of Salzburg, Salzburg	PT 2.5 yrs / 5 semesters	90	10-15	a,b,c	public
	Bachelor Sign Language Interpreting	Health University of Applied Sciences Tyrol, Innsbruck	FT 3 yrs	180	10-16	a,c	public
	BA Dolmetschen: Österreichische Gebärdensprache - Deutsch	University of Applied Sciences Kärnten, Klagenfurt	FT 3 yrs	180	8-13	b,c[28]	public
Belgium-Flanders	Toegepaste Taalkunde - Master in het tolken	KU Leuven, Antwerp	5 yrs	260	~25	a,b,c	private
Belgium-Wallonia	Master en interprétation	UCLouvain, Louvain-la-Neuve	FT 5 yrs	120	11	a,b,c	private
Czech Republic	Jazyky a komunikace neslyšících (Languages and communication of the Deaf)[29]	Institute of Czech and Deaf Studies Faculty of Arts (Charles University), Prague	FT 3 yrs	180	20	b,c	public
	Tlumočnictví českého znakového jazyka (Czech Sign Language Interpreting program)	Vyšší odborná škola, Střední škola, Základní škola a Mateřská škola, Hradec Králové, Štefánikova 549	FT 3 years	180		a	public
Denmark	Bachelor's degree in sign language interpreting	Københavns Professionshøjskole, Copenhagen	FT 3.5 yrs	210	20	a,b,c	public
England, W & NI	BSL/English interpreting – BA[30] & MA[31]	University of Wolverhampton, Wolverhampton	BA: PT 6 yrs, FT 3 yrs MA: PT 3 yrs		DK	b	public

[26] (a) Officially supervised & recognized by the government, (b) Recognized or supported by the sign language interpreter association, (c) Recognized or supported by the national deaf association, (d) Other
[27] The BA focuses on language acquisition, cultural knowledge and basic competences in translation
[28] This program started in October 2025, not clear yet if it is recognized by the government
[29] Not officially recognized for SLI training, but well known, has accreditation
[30] https://www.wlv.ac.uk/courses/ba-hons-british-sign-language-interpreting/
[31] https://www.wolverhampton.ac.uk/courses/ma-interpreting/

Table 4.3 Overview of sign language interpreter training programs in Europe

Country / region	Program	Institute, City	Years and FT/PT	ECTS	# students enrolled	Status[26]	Public/ private
	BSL/English interpreting and Translation – postgraduate[32]	University of Central Lancashire, Preston	PT[33] 2-3 yrs				public
Finland	Tulkki (AMK), Viittomakielen ja puhevammaisten tulkkaus	Diaconia University of Applied Sciences (Diak), Helsinki	FT/PT 4 yrs	240	DK	a,b,c	public
	Tulkki (YAMK), Tulkkaustoiminnan kehittäminen, Master's Degree Program in the Development of Interpreting Practices	Joint program of Diaconia University of Applied Sciences (Diak) and Humak University of Applied Sciences (Humak), Helsinki	FT/PT 1.5-2 yrs	90	DK	a,b,c	public
	Tulkki (AMK), viittomakielen tulkkaus	Humak University of Applied Sciences	FT/PT 4 yrs	240	DK	a,b,c	public
	European Master in Sign Language interpreting		PT 2.5 yrs	90	DK		
	Oikeustulkkauksen erikoisammattitutkinto	Tampereen Aikuiskoulutuskeskus TAKK, Tampere Adult Education Centre, TAKK[34] Tampere	PT 1.5 yrs	NA	DK	a,b	Private
	Viittomakielen asiantuntija (bachelor and masters), *offers MA studies in sign language*	Jyväskylän yliopisto, Jyväskylä	FT BA: 3 yrs MA: 2 yrs	240 / 120	DK	a,b,c	Public
France	Master Interprétariat langue des signes française/français	Université de Lille, Lille	FT 2 yrs	*[35]	DK	b,c	public
	Master Interpretation LSF	ESIT Université Paris 3 Paris					
	Master Interprétation Langue des Signes Française / Français	Université Paris 8, Paris					
	Master parcours Traduction Interprétation et Médiation Linguistique	D-TIM - University of Toulouse 2 Jean Jaurès Toulouse					
	Master Sciences du Langage, Interpretariat en Langue des Signes	Université de Rouen Rouen					
Georgia	Three level modules		3 months		DK	a,c	public

[32] https://www.lancashire.ac.uk/postgraduate/courses/british-sign-languageenglish-interpreting-pgdipma

[33] The course is taught over weekends, one weekend every other month for two years.

[34] Not offered every year.

[35] This program trains future hearing interpreters, Deaf translators and Deaf intermediators. Regarding the topics students are educated on, it is important to keep in mind that University does not allow unlimited funding, as a result the amount of training hours is rather low.

Table 4.3 Overview of sign language interpreter training programs in Europe

Country / region	Program	Institute, City	Years and FT/PT	ECTS	# students enrolled	Status[26]	Public/ private
Germany	Modularisierter Diplomstudiengang: Diplom-Gebärdensprachdolmetscher/in (FH)	Westsächsische Hochschule Zwickau (FH), Zwickau	FT 4 yrs	240	DK	b,c	public
	Gebärdensprachdolmetschen – BA	Hochschule Magdeburg Stendal, Magdeburg	FT 4 yrs	DK	DK	a,b,c	public
	European Master in Sign Language interpreting		PT 2.5 yrs	90	DK		
	Gebärdensprachdolmetschen – BA[36]	Universität Hamburg, Hamburg	FT 3.5 yrs	DK		b,c	public
	Gebärdensprachdolmetschen – MA[37]						
	Gebärdensprachdolmetschen – BA	Hochschule für angewandte Wissenschaften Landshut, Landshut	FT 3.5 yrs	210	DK	b,c	public
	Dolmetschen für Deutsche Gebärdensprache – BA	Universität zu Köln, Cologne	FT 3.5 yrs	210	DK		
	Gebärdensprachdolmetschen / Dolmetschen und Übersetzen für Deutsche Gebärdensprache (MA)	Humboldt University, Berlin	FT 2 yrs	DK	DK	b,c	public
	Gebärdensprachdolmetschen – BA	Hochschule Fresenius, Idstein					
	Gebärdensprachdolmetschen – BA	Pädagogische Hochschule Heidelberg, Heidelberg	FT 3.5 yrs	210	DK		public
Greece	Examinations for candidate Greek Sign Language interpreters (interpreter training provided by sign language schools)	The Association of Greek Sign Language Interpreters (SDENG), in collaboration with the Hellenic Federation of the Deaf and under the auspices of the Ionian University, Athens and Thessaloniki	PT		DK	b,c	private
Hungary	Sign language interpreter (hearing)	SINOSZ, Budapest	PT 2 yrs 1000 hrs	NA	14	a,b,c	public
	Sign language interpreter (deaf)						
Iceland	Táknmálsfræði og táknmálstúlkun (Sign Language studies and interpretation)	University of Iceland, Reykjavik	PT 3 yrs	180	Annually 10 new students	a	public

[36] https://www.slm.uni-hamburg.de/studium/studiengaenge/ba-studiengaenge/gebaerdensprachdolmetschen.html
[37] https://www.slm.uni-hamburg.de/studium/studiengaenge/ma-studiengaenge/gebaerdensprachdolmetschen.html

Table 4.3 Overview of sign language interpreter training programs in Europe							
Country / region	**Program**	**Institute, City**	**Years and FT/PT**	**ECTS**	**# students enrolled**	**Status[26]**	**Public/ private**
Ireland	Bachelor of Deaf Studies	Centre for Deaf Studies, Trinity College Dublin, Dublin	FT 4 yrs		DK	a	public
Italy	Corso sperimentale ad orientamento professionale "Interprete in LIS e LIST"	Università di Parma, Parma	FT 3 yrs	180	DK	a	public
	Master in Teorie e Tecniche di Traduzione e Interpretazione Italiano-LIS / LIS-Italiano	Università di Catania / Struttura Didattica Speciale Lingue e Letterature della sede di Ragusa, Catania / Ragusa	FT 1 yr	120	DK	a, b	public
	CdL in Interpretariato e traduzione in lingua dei segni italiana (LIS) e lingua dei segni italiana tattile (LIST)	Interateneo Università degli Studi di Milano Bicocca, Milan	FT 3 yrs	180	DK	a	public
	Comunicazione e Interpretariato in Lingua dei segni italiana (LIS e LIST)	Sapienza Università di Roma, Rome	FT 3 yrs	180	DK	a	public
	Comunicazione e Interpretariato in Lingua dei segni italiana OR Interpretariato e traduzione in lingua dei segni Italiana (LIS) e lingua dei segni Italiana tattile (LIST)	Sapienza, Università di Roma | Università degli studi di Milano Bicocca, Rome, Milan, Parma					
	Corso Interpretazione-Traduzione LIS/Italiano (biennale)	Gruppo SILIS (con patrocini) Rome	FT 2 yrs		DK	a,b	private
Latvia	Surdotulks (Sign Language Interpreter)	Sociālās integrācijas valsts aģentūra (The Social Integration State Agency), Jūrmala	FT 2 yrs		12	a	public
Lithuania	Lietuvių gestų kalbos vertimas	Vilniaus kolegija, Vilnius	FT 3 yrs	180	DK	a	public
Netherlands	Tolk Nederlandse Gebarentaal-NL	University of Applied Sciences Utrecht, Utrecht	FT/PT 4 yrs	240	DK	a	public
Norway	Tegnspråk og tolking	Høgskulen på Vestlandet (HVL), Bergen	FT 3 yrs	180	DK	a	public
	Norsk tegnspråk	NTNU, Trondheim	FT 3 yrs	180	DK	a	public
	Norsk tegnspråk	Oslo Metropolitan University, Oslo	FT 3 yrs	180	DK	a	public

Table 4.3 Overview of sign language interpreter training programs in Europe

Country / region	Program	Institute, City	Years and FT/PT	ECTS	# students enrolled	Status[26]	Public/ private
Poland	MA specialization in Polish Sign Language (PJM) within Translation Studies[38]	University of Warsaw, Institute of Applied Linguistics (ILS UW), Warsaw	FT 2 yrs	120	2	note[39]	public
Portugal	Tradução e Interpretação de Língua Gestual Portuguesa	Instituto Politécnico de Setúbal - Escola Superior de Educação, Setúbal	FT 3 yrs	180	40	a,b,c	public
	Licenciatura em Língua Gestual Portuguesa	Instituto Politécnico de Coimbra - Escola Superior de Educação, Coimbra	FT 3 yrs	180	100	a,b,c	public
	Tradução e Interpretação em Língua Gestual Portuguesa	Instituto Politécnico do Porto, Porto	FT 3 yrs	180	DK	a,b,c	public
Romania	Curs de Comunicare in limba semnelor romane[40]	Facultatea de Jurnalism, University of Bucharest, Bucharest	PT on weekends : 8 months (228 hours)	NA	25-30		public
Russia	Theory and practice of sign language interpreting in intercultural communication (RSL, English)	Moscow State Linguistic University, Moscow	FT 4 yrs		40[41]	a, b	public
	Linguistics, 'English and RSL Interpreter'[42]	Novosibirsk State Technical University, Novosibirsk	FT 4 yrs		23 annually	a, b	public

[38] This is a very recent development. In 2025, Aleksandra Kalata-Zawłocka launched the first university-level language specialization focused primarily on Polish Sign Language (PJM) at the Institute of Applied Linguistics, University of Warsaw (ILS UW). The program is part of a broader translation studies master's track, which means that recruitment follows the same rules as for spoken-language specializations. The main challenge is that candidates are required to demonstrate PJM proficiency at the C1 level upon entry to this two-year MA program, yet no official C1-level PJM examination exists in Poland. As a result, only a very small group of applicants qualified: those who had previously completed a master's degree in PJM philology or a BA in translation at ILS with PJM as an additional language (language C). This makes the current candidate pool extremely limited. The hope is that in the coming years the University of Warsaw will establish formal PJM proficiency exams, allowing a broader range of Deaf and hearing candidates to apply. Despite these constraints, the launch of this specialization in October 2025 marks a major step forward—it is currently the only such program in Poland.

[39] Officially supervised & recognized by University of Warsaw. Not "by the government" because there is no law on country level which would consider SLI as a profession anyway.

[40] Participants receive a certification of professional competence. Advertised as a post-graduate degree course during weekends (228 hours of theory and practice): https://fjsc.unibuc.ro/comunicare-si-interpretare-in-limba-semnelor-romane/

[41] Annually 10 new students enroll. Current cohort is therefore 40 students (4-year program).

[42] Key subjects in the curriculum: - First foreign language (English) - A second foreign language (Russian Sign Language) - Russian language and speech culture - Basics of linguistics - Theory and practice of intercultural communication - Theory and practice of translation - Russian Sign Language as a linguistic system - History and culture of deaf people - Sign language pedagogy and psychology - Medical and social expertise in rehabilitation work with deaf people -Professional ethics of Russian Sign Language interpreters - Practical translation course - Culture of communication in Russian Sign Language - Interpreting for Deafblind - International Sign Communication The implementation of this educational program started at NSTU in 2013.

Table 4.3 Overview of sign language interpreter training programs in Europe							
Country / region	**Program**	**Institute, City**	**Years and FT/PT**	**ECTS**	**# students enrolled**	**Status[26]**	**Public/ private**
	Linguistics, profile 'Theory and practice of Russian sign language interpreting'[43]	Kazan Federal University, Kazan	FT / PT 4.5 yrs		DK	a, b	public
Scotland	MA British Sign Language (Interpreting, Translation & Applied Language Studies)	Heriot Watt University, Edinburgh	FT 4 yrs		DK	a, b, c	public
	Advancing Practice in Signed/Spoken Language Interpreting - MSc	Queen Margaret University, Edinburgh	PT 2-7 yrs				public
Slovakia	Slovenský jazyk v komunikácii nepočujúcich	Trnava University, Trnava	FT 4 yrs		27 (24 hearing, 3 deaf)	a, b, c	public
Spain	Comunidad Sorda y Lengua de Signos	Universidad Rey Juan Carlos, Madrid	FT 4 yrs	240	DK	a, b, c	public
	Traducción e Interpretación (opción de especialidad en LSC)[44]	Universitat Pompeu Fabra, Barcelona	FT 4 yrs	240	DK		public
Sweden	Teckenspråks- och dövblindtolk (Sign language- and deafblind interpreter)	Fellingsbro Folkhögskola, Örebro	FT 4 yrs		DK	a, b, c	public
	Teckenspråks- och dövblindtolk (Sign language- and deafblind interpreter)	Västanviks Folkhögskola, Leksand	FT 4 years		DK	a, b, c	public
	Kandidatprogram i teckenspråk och tolkning, inriktning Tolkning mellan svenskt teckenspråk och svenska	Stockholm University, Stockholm	FT 3 yrs	180		a, b, c	public
	Folkhögskola Tolkutbildning för döva (Interpreting program for deaf persons)[45]	Västanviks, Distant studies & 10 x onsite	2 yrs				
Switz.-Ger.	BA of Arts in Sign Language Interpreting[46]	Interkantonale Hochschule für Heilpädagogik, Zürich	8-12 semesters	180	17	a, b, c	public
Switz.-It.	Bachelor	University of Geneva Geneva	FT 3 yrs	180	3	a,b	public

[43] Qualification – Bachelor of Linguistics. The form of education is full-time and part-time. The term of study based on basic general education is 4 years and 6 months. Key subjects in the curriculum: - General linguistics; - General theory of translation; - Theory of sign language translation; - Practical Russian language course; - Practical course of Russian Sign Language ; - Theoretical and practical foreign language course; - Russian Sign Language as a linguistic system; -Psycholinguistic aspects of learning sign language; - Legal status of persons with disabilities; - History and culture of Russian Sign Language ; - Professional ethics of Russian Sign Language interpreter; Workshop on the communication in Russian Sign Language ; Workshop on the written translation culture from verbal Russian into Russian Sign Language . Russian translation workshop on written translation culture. Sign language to verbal Russian; Russian oral translation culture workshop on Russian Sign Language into verbal Russian; A workshop on the culture of oral translation from verbal Russian language to Russian Sign Language; International Sign. The implementation of this educational program started at KFU in 2021.

[44] https://issuu.com/universitatpompeufabra/docs/grau_en_traducci_i_interpretaci_

[45] https://www.folkhogskola.nu/sok-skolor/Dalarnas-lan/Vastanviks-folkhogskola/Kurser/2024ht/tolkutbildning-for-dova/

[46] https://www.hfh.ch/ausbildung/ba-gebaerdensprachdolmetschen/auf-einen-blick

Table 4.3 Overview of sign language interpreter training programs in Europe							
Country / region	**Program**	**Institute, City**	**Years and FT/PT**	**ECTS**	**# students enrolled**	**Status[26]**	**Public/ private**
Turkey	Department of Sign Language and Deaf Studies – master's program in Sign Language Interpreting	Ankara University Institute of Social Sciences, Ankara	PT 2 yrs	120[47]	DK[48]	a	public
Ukraine[HI]	Sign Language Interpreters Training Course	Collective Enterprise 'Educational and Rehabilitation Center of the Ukrainian Society of the Deaf', Kyiv	FT 20 weeks		14	a,c	private
	Social work and counseling, sign language interpreter	Kyiv College of Applied Sciences, Kyiv	FT 2 yrs	120		a,c	public
	Social work and counseling, sign language interpretation	Poltava Vocational College of Ukraine University, Poltava	FT 2-3 years	120	DK	a	

DK=Don't know

[47] Thesis-based master's programs consist of at least seven courses, a seminar, other learning activities, and a thesis, with a minimum of 21 national credits and 120 ECTS credits. Of the total 120 ECTS credits, 60 ECTS are allocated to courses and the seminar, while the remaining 60 ECTS are assigned to thesis-related specialization courses. https://www.ankara.edu.tr/programlar/3/903/4317-1939#access_further

[48] However, since its establishment, the program has admitted approximately 15 students. But the program is not currently admitting students.

Table 4.4 Sign language interpreter training programs and deaf students

Country / region	Enrollment possible for deaf students?	Requirements specific for deaf students	Curriculum adapted to educate deaf and hearing students in the same program	Specific modules for deaf and hearing students?
Austria				
University of Graz (ITAT)[49]	yes[50]	DK	yes	DK
GESDO, Linz	no			
MODUS,[51] University of Salzburg	yes	no	yes	DK
Health University of Applied Sciences Tyrol, Innsbruck	no			
University of Applied Sciences Kärnten, Klagenfurt	no			
Belgium - Flanders				
KU Leuven, Antwerp	yes	DK	no	no
Belgium - Wallonia				
UC Louvain, Louvain-la-Neuve	no		no	
Czech Republic				
Charles University, Prague	yes	no	no	no
Hradec Králové	~[52]			no
Københavns Professionshøjskole, Copenhagen	yes	DK	no	no
England, W & NI				
University of Wolverhampton, Wolverhampton	yes		DK	DK
University of Central Lancashire, Preston				
Finland				
Diaconia University of Applied Sciences (Diak) Helsinki	DK		DK	DK
Joint program of Diaconia University of Applied Sciences (Diak) and Humak University of Applied Sciences (Humak), Helsinki[53]	yes		DK	DK
Humak University of Applied Sciences, Helsinki: Tulkki (AMK), viittomakielen tulkkaus European Master in Sign Language interpreting[54]	1.DK 2. yes	DK	DK	DK

[49] Austrian Sign Language (ÖGS) is only one of 12 languages that is offered at this program; it is possible to study 3 or even 4 working languages; for example, with German as the first language, it is possible to study ÖGS and English and even add French e.g.; although the program does not offer specific courses to study interpreting from EN to ÖGS or vice versa, there are additional events and learning environments organized where students can test this combination out for themselves.

[50] Deaf students can attend the translation track and the courses.

[51] This curriculum was called "Logo" before and was only meant for educating deaf students; the curriculum has been adapted and since 2019 it is meant for deaf and hearing students.

[52] In theory yes, in practice not.

[53] Program in Finnish, and common program for SLIs and interpreters of spoken languages (community interpreting).

[54] www.eumasli.eu

Table 4.4 Sign language interpreter training programs and deaf students				
Country / region	**Enrollment possible for deaf students?**	**Requirements specific for deaf students**	**Curriculum adapted to educate deaf and hearing students in the same program**	**Specific modules for deaf and hearing students?**
Tampereen Aikuiskoulutuskeskus TAKK, Tampere Adult Education Centre, TAKK, Tampere	DK	DK	DK	DK
Jyväskylän yliopisto, Jyväskylä	yes	DK	DK	DK
France				
Université de Lille, Lille	yes	No	DK	yes
ESIT Université Paris 3, Paris	no			
Université Paris 8, Paris				
D-TIM - University of Toulouse 2 Jean Jaurès, Toulouse				
Université de Rouen, Rouen				
Georgia				
Three level modules	no			
Germany				
Westsächsische Hochschule Zwickau (FH), Zwickau	DK	DK	DK	DK
Hochschule Magdeburg Stendal, Magdeburg: 1. Gebärdensprachdolmetschen – BA 2. European Master in Sign Language interpreting	1. DK 2. yes			DK
Universität Hamburg, Hamburg: 1. Gebärdensprachdolmetschen – BA 2. Gebärdensprachdolmetschen – MA	DK			
Hochschule für angewandte Wissenschaften Landshut, Landshut	DK			
Universität zu Köln, Cologne	DK			
Humboldt University, Berlin	DK			
Hochschule Fresenius, Idstein	DK			
Pädagogische Hochschule Heidelberg, Heidelberg	DK			
Greece				
The Association of Greek Sign Language Interpreters (SDENG), in collaboration with the Hellenic Federation of the Deaf and under the auspices of the Ionian University, Athens and Thessaloniki	no			
Hungary				
SINOSZ, Budapest 1.Sign language interpreter (hearing) 2.Sign language interpreter (deaf)	no/yes[55]			

[55] Deaf students have their own program which partly runs together with the hearing students.

Table 4.4 Sign language interpreter training programs and deaf students				
Country / region	**Enrollment possible for deaf students?**	**Requirements specific for deaf students**	**Curriculum adapted to educate deaf and hearing students in the same program**	**Specific modules for deaf and hearing students?**
Iceland				
University of Iceland, Reykjavik	yes	no	yes[56]	Interpreting into spoken language is only for hearing students
Ireland				
Centre for Deaf Studies, Trinity College Dublin, Dublin	yes	DK	DK	DK
Italy				
Università di Parma, Parma	Yes	DK	DK	DK
Università di Catania / Struttura Didattica Speciale Lingue e Letterature della sede di Ragusa, Catania / Ragusa	DK			
interateneo Università degli Studi di Milano Bicocca, Milan				
Sapienza Università di Roma, Rome	Yes	DK	DK	DK
Sapienza, Università di Roma | Università degli studi, Rome, Milan, Parma di Milano Bicocca	Yes	DK	DK	
Gruppo SILIS (con patrocini), Rome	DK		DK	
Latvia				
Sociālās integrācijas valsts aģentūra (The Social Integration State Agency), Jūrmala	yes[57]	no	yes	no
Lithuania				
Vilniaus kolegija, Vilnius	yes	DK	DK	no
Netherlands				
University of Applied Sciences Utrecht, Utrecht	yes[58]			yes[59]
Norway				
Høgskulen på Vestlandet (HVL), Bergen	yes		DK	DK

[56] The assignments are adapted, and the teaching is mostly in Sign Language - interpreted if there are deaf student when taught in spoken Icelandic.

[57] All lectures are interpreted into SL.

[58] The educational institution (ICG) for Dutch Sign Language interpreters was in the process of developing a graduation track for Deaf interpreters. For various reasons, the ICG has decided not to continue developing this graduation track in the program. In the accreditation panel's advisory report on the program, the accessibility of the education for deaf students is identified as a point of attention. The NBTG will have a meeting with the training program on this topic. Based on that, the NBTG will determine any possible next steps.

[59] After many years of consultation and preparation by a working group of Deaf interpreters and the NBTG, a framework for a graduation track for Deaf interpreters was developed in collaboration with IGT&D. The aim of this graduation track was to further professionalize the profession and, by graduating Deaf interpreters, to meet the needs of a large group of interpreter users. It had been agreed with the professional register (RTGS) that a first group of Deaf interpreters who meet certain conditions—such as completing an assessment and following the Dutch Sign Language Interpreter training—would be provisionally registered in the register. This allowed these interpreters to work within the government-funded interpreting provision.

Table 4.4 Sign language interpreter training programs and deaf students

Country / region	Enrollment possible for deaf students?	Requirements specific for deaf students	Curriculum adapted to educate deaf and hearing students in the same program	Specific modules for deaf and hearing students?
NTNU, Trondheim	yes	DK	DK	DK
Oslo Metropolitan University, Oslo	yes	DK	DK	DK
Poland				
University of Warsaw, Institute of Applied Linguistics (ILS UW), Warsaw	yes[60]	No	yes	no
Portugal				
Instituto Politécnico de Setúbal - Escola Superior de Educação, Setúbal	no		no	no
Instituto Politécnico de Coimbra - Escola Superior de Educação, Coimbra	yes		no	no
Instituto Politécnico do Porto, Porto	no			no
Romania				
Facultatea de Jurnalism, University of Bucharest, Bucharest	DK[61]			
Russia				
Moscow State Linguistic University, Moscow	yes	no	no	no
Novosibirsk State Technical University, Novosibirsk	DK			no
Kazan Federal University, Kazan	DK	No		
Scotland				
Heriot Watt University, Edinburgh	DK			
Queen Margaret University, Edinburgh	DK			
Slovakia				
Trnava University, Trnava	yes	no	Yes	no
Spain				
Universidad Rey Juan Carlos, Madrid	yes	no	no	no
Traducción e Interpretación (opción de especialidad en LSC), Barcelona				
Sweden				
Fellingsbro Folkhögskola, Örebro	no			
Västanviks Folkhögskola, Leksand	no		no	
Stockholm University, Stockholm	yes[62]	no	no	yes
Västanviks (Distant + onsite)	yes[63]			

60 Most of subjects are led by deaf-hearing team.
61 The course is in Romanian.
62 There are two branches: intramodal interpreting and interpreting between Swedish Sign Language and Swedish.
63 This is a dedicated program for deaf interpreters.

Table 4.4 Sign language interpreter training programs and deaf students				
Country / region	**Enrollment possible for deaf students?**	**Requirements specific for deaf students**	**Curriculum adapted to educate deaf and hearing students in the same program**	**Specific modules for deaf and hearing students?**
		Switzerland-German region		
Interkantonale Hochschule für Heilpädagogik, Zürich	no		no	no
		Switzerland-Italian region		
University of Geneva, Geneva	yes	no	no	no
		Turkey		
Ankara University Institute of Social Sciences, Ankara	yes	no	Yes	no
		Ukraine		
Collective Enterprise 'Educational and Rehabilitation Center of the Ukrainian Society of the Deaf', Kyiv	yes	no	no	no
Kyiv College of Applied Sciences, Kyiv	yes		no	no
Poltava Vocational College of Ukraine University, Poltava	yes	no	no	no

DK= Don't know

4.4 Prerequisites to attend the program

Sign language interpreter training programs have prerequisites for students to apply. Figure 4.2 gives a total overview of the type of prerequisites, with a high school diploma being the main prerequisite followed an entry exam. A required minimum level of language skills is sometimes part of the entry exam, or it is a separate criterion.

Fig. 4.2 Prerequisites to attend the educational program for SLI (n=65)

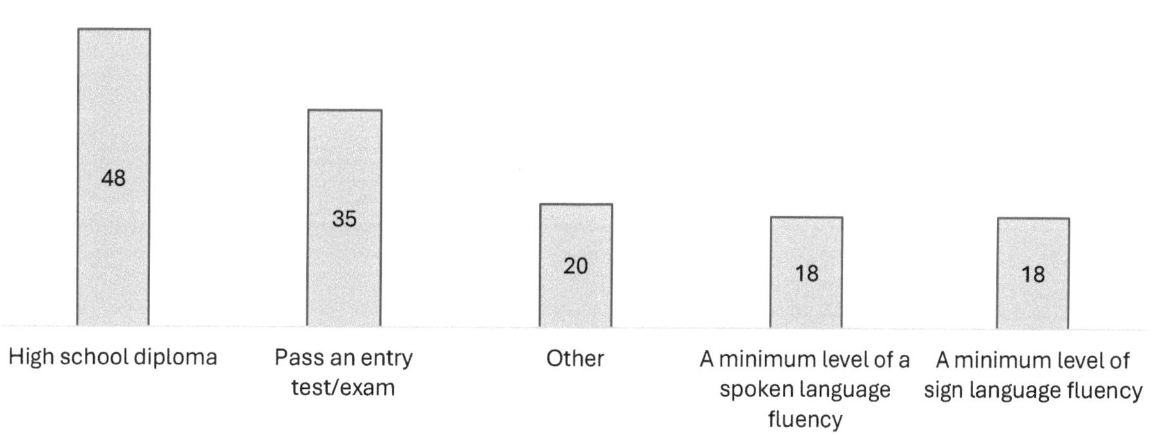

Table 4.5 details the requirements further of the prerequisites to attend an educational program. Additional details were added, such as hyperlinks and informative collected from the website of the programs.

Table 4.5 Prerequisites to attend educational programs

Country / region	**Prerequisites** HS = High School diploma EE= Entry Exam SpL= min. level of Spoken Language fluency SL = min. level of Sign Language fluency OTH = Other Ø = None	**CEFR**
	Austria	
University of Graz (ITAT)[64]	HS, SpL, SL (A2[65])	yes
GESDO, Linz	HS, EE	DK
MODUS, University of Salzburg	EE SpL & SL: C1 if it is your 1st language and B2 if it is your 2nd language	yes
Health University of Applied Sciences Tyrol, Innsbruck	HS, EE, SpL, SL (A1)	yes
University of Applied Sciences Kärnten, Klagenfurt	HS, SpL, EE(?)[66]	yes
	Belgium-Flanders	
KU Leuven, Antwerp	HS *Note*: During the Bachelor you must choose French, English or German. In the Master you can choose to only continue with VGT or combine with the other language.	yes
	Belgium-Wallonia	
UCLouvain, Louvain-la-Neuve	EE, SpL, SL (B2)	
	Czech Republic	
Charles University, Prague	HS, EE *Note*: as university program, the students need to pass a test; for SLI they choose the module after a year of their study, need to have sufficient skills in SL (based on SL test).	yes
Hradec Králové	HS, EE, SpL OTH: medical checkup that you can do this program and job, motivational letter, CV.	
	Denmark	
Københavns Professionshøjskole, Copenhagen	HS, EE, SpL	yes
	England, W & NI	
University of Wolverhampton, Wolverhampton	BA: HS; OTH[67] MA: SpL, SL (C1: BSL & English)	DK

64 Austrian Sign Language (ÖGS) is only one of 12 languages that is offered at this program; it is possible to study 3 or even 4 working languages; for example, with German as the first language, it is possible to study ÖGS and English and even add French e.g.; although the program does not offer specific courses to study interpreting from EN to ÖGS or vice versa, there are additional events and learning environments organized where students can test this combination out for themselves.

65 The University also offers a one-year course to get to this level, so students can start their studies and attend this course parallel.

66 Not sure if there is an entry exam. The program started in October 2025.

67 https://www.wlv.ac.uk/courses/ba-hons-british-sign-language-interpreting/

Table 4.5 Prerequisites to attend educational programs		
Country / region	**Prerequisites** HS = High School diploma EE= Entry Exam SpL= min. level of Spoken Language fluency SL = min. level of Sign Language fluency OTH = Other Ø = None	**CEFR**
	OTH: A reference from a tutor who has supervised the applicant's undergraduate studies and/or employer; A first or second class (2:2 minimum) undergraduate honors degree from a UK university or overseas equivalent, or a professional qualification and/or experience).	
University of Central Lancashire, Preston		DK
Finland		
Diaconia University of Applied Sciences (Diak) Helsinki	HS, EE	DK
Joint program of Diaconia University of Applied Sciences (Diak) and Humak University of Applied Sciences (Humak), Helsinki	EE OTH: Bachelor's degree and 2 years of working experience.	DK
Humak University of Applied Sciences, Helsinki: Tulkki (AMK), viittomakielen tulkkaus European Master in Sign Language interpreting	HS HS, BA	1.yes 2. yes
Tampereen Aikuiskoulutuskeskus TAKK, Tampere Adult Education Centre, TAKK, Tampere	OTH: no information.	
Jyväskylän yliopisto, Jyväskylä	HS, EE	DK
France		
Université de Lille, Lille	EE, SpL, SL (B2/C1) OTH: BA diploma	yes
ESIT Université Paris 3, Paris	HS, EE, SpL, SL	
Université Paris 8, Paris	HS, EE, SpL, SL	
D-TIM - University of Toulouse 2 Jean Jaurès, Toulouse	OTH: English-French-French Sign Language Translation-Mediation Licence from DeTIM, aSDL Licence or an equivalent bac+3 qualification.	
Université de Rouen, Rouen	EE,[68] SpL, SL,	
Georgia		
Three level modules	SL	
Germany		
Westsächsische Hochschule Zwickau (FH), Zwickau[69]	HS	DK
Hochschule Magdeburg Stendal, Magdeburg: 1. Gebärdensprachdolmetschen – BA [70] 2. European Master in Sign Language interpreting[71]	1. HS, OTH: 60 hours of DGS 2. see www.eumasli.eu	
Universität Hamburg, Hamburg: 1. Gebärdensprachdolmetschen – BA	1. HS, OTH.[72]	DK

[68] https://lsh.univ-rouen.fr/intepretariat-en-langues-des-signes-823224.kjsp
[69] https://www.whz.de/fileadmin/studieninteressenten/docs/Flyer/20250909_Gebaerdensprachdolmetschen.pdf
[70] https://studieren.h2.de/studiengaenge/bachelor/gebaerdensprachdolmetschen#c766
[71] https://eumasli.humak.fi/application/
[72] https://www.uni-hamburg.de/campuscenter/bewerbung/bachelor-staatsexamen/zugangsvoraussetzungen.html

Table 4.5 Prerequisites to attend educational programs

Country / region	**Prerequisites** HS = High School diploma EE= Entry Exam SpL= min. level of Spoken Language fluency SL = min. level of Sign Language fluency OTH = Other Ø = None	**CEFR**
2. Gebärdensprachdolmetschen – MA	2. HS, OTH.[73]	
Hochschule für angewandte Wissenschaften Landshut, Landshut[74]	HS, SpL (B2 spoken German) SL (A1)	DK
Universität zu Köln, Cologne[75]	HS	
Humboldt University, Berlin	HS, SL (B2) OTH.[76]	yes
Hochschule Fresenius, Idstein[77]	HS, SpL (German: excellent), SL (A1) OTH: "to be able to see and hear"	
Pädagogische Hochschule Heidelberg, Heidelberg	HS	DK
Greece		
The Association of Greek Sign Language Interpreters (SDENG), in collaboration with the Hellenic Federation of the Deaf and under the auspices of the Ionian University, Athens and Thessaloniki	HS	
Hungary		
SINOSZ, Budapest 1.Sign language interpreter (hearing) 2.Sign language interpreter (deaf)	HS, EE, SpL, SL (B1)	yes
Iceland		
University of Iceland, Reykjavik	HS	yes
Ireland		
Centre for Deaf Studies, Trinity College Dublin, Dublin	EE Note: The exam in Ireland is called the 'Leaving Cert' and is like a high school diploma. Students can also enter through the mature student access route.	yes
Italy		
Università di Parma, Parma	HS, SL (B1)	yes
Università di Catania / Struttura Didattica Speciale Lingue e Letterature della sede di Ragusa, Catania / Ragusa	SL (B2) OTH: Bachelor's degree	yes
interateneo Università degli Studi di Milano Bicocca, Milan	EE	yes
Sapienza Università di Roma, Rome	EE	yes
Sapienza, Università di Roma | Università degli studi, Rome, Milan, Parma di Milano Bicocca		
Gruppo SILIS (con patrocini), Rome	EE	

[73] https://www.slm.uni-hamburg.de/studium/studiengaenge/ma-studiengaenge/gebaerdensprachdolmetschen.html
[74] https://www.haw-landshut.de/static/Fakultaet_GKM/Dateien/Studiengangsflyer/GKM_GSD-Studiengangsflyer-IDS-6-Seiter.pdf
[75] https://studieninteressierte.uni-koeln.de/studienangebot/index_ger.html?app=true&id=47
[76] https://www.hu-berlin.de/fileadmin/Dokumentenpool/Studium/202504_WiSe2025-26_zzr/2.2.1.62.-2024-D%C3%9CDGS.pdf
[77] https://www.hs-fresenius.de/berufsbegleitend/bachelor/gebaerdensprachdolmetscher/

Table 4.5 Prerequisites to attend educational programs		
Country / region	**Prerequisites** HS = High School diploma EE= Entry Exam SpL= min. level of Spoken Language fluency SL = min. level of Sign Language fluency OTH = Other Ø = None	**CEFR**
Latvia		
Sociālās integrācijas valsts aģentūra (The Social Integration State Agency), Jūrmala	HS, EE	DK
Lithuania		
Vilniaus kolegija, Vilnius	HS	DK
Netherlands		
University of Applied Sciences Utrecht, Utrecht	HS	yes
Norway		
Høgskulen på Vestlandet (HVL), Bergen	HS	
NTNU, Trondheim	HS	
Oslo Metropolitan University, Oslo	HS, EE, SL	DK
Poland		
University of Warsaw, Institute of Applied Linguistics (ILS UW), Warsaw	SL (C1)	yes
Portugal		
Instituto Politécnico de Setúbal - Escola Superior de Educação, Setúbal	HS	no
Instituto Politécnico de Coimbra - Escola Superior de Educação, Coimbra	HS, EE	no
Instituto Politécnico do Porto, Porto	HS, EE	no
Romania		
Facultatea de Jurnalism, University of Bucharest, Bucharest	OTH: University degree diploma.	DK
Russia		
Moscow State Linguistic University, Moscow	HS, EE OTH: passing final state exams in Russian language, history and math after graduating school.	yes
Novosibirsk State Technical University, Novosibirsk	HS, EE	no
Kazan Federal University,Kazan	HS, EE	
Scotland		
Heriot Watt University, Edinburgh[78]	HS, SpL: Minimum English level IELTS 6.5	
Queen Margaret University, Edinburgh[79]	OTH: Up-to-date registration for interpreters from countries where registration is available. A minimum of three years of experience post-registration, or at least three years of experience for those for whom registration is not available.	

[78] https://www.hw.ac.uk/study/undergraduate/british-sign-language-interpreting-translating-and-applied-language
[79] https://www.qmu.ac.uk/study-here/postgraduate-study/2026/msc-advancing-practice-in-signed-spoken-language-interpreting

Table 4.5 Prerequisites to attend educational programs		
Country / region	**Prerequisites** HS = High School diploma EE= Entry Exam SpL= min. level of Spoken Language fluency SL = min. level of Sign Language fluency OTH = Other Ø = None	**CEFR**
Slovakia		
Trnava University, Trnava	HS, EE	yes
Spain		
Universidad Rey Juan Carlos, Madrid[80]	HS, OTH: No information.	yes
Traducción e Interpretación (opción de especialidad en LSC), Barcelona		
Sweden		
Fellingsbro Folkhögskola, Örebro	HS, SpL	DK
Västanviks Folkhögskola, Leksand	HS, SpL	DK
Stockholm University, Stockholm	HS, SpL OTH: Swedish upper secondary school courses English B and Social Studies A.	DK
Västanviks (Distant + onsite)		
Switzerland-German region		
Interkantonale Hochschule für Heilpädagogik, Zürich	HS, EE, SpL	yes
Switzerland-Italian region		
University of Geneva, Geneva	HS, EE, SpL, SL (B2)	yes
Turkey		
Ankara University Institute of Social Sciences, Ankara	EE, SpL, SL (B1) OTH: a university degree.	yes
Ukraine		
Collective Enterprise 'Educational and Rehabilitation Center of the Ukrainian Society of the Deaf', Kyiv	HS, EE, SL (min. elementary)	no
Kyiv College of Applied Sciences, Kyiv	HS, EE	no
Poltava Vocational College of Ukraine University, Poltava	HS, EE	no

DK=Don't know

4.5 Interpreting for deafblind persons

Fifteen countries and regions offer programs on interpreting for deafblind persons (Table 4.6). In Belgium-Wallonia, Czech Republic, England W&NI, Georgia, Greece, Latvia, Switzerland-It, Turkey, and Ukraine have not such programs. All other respondents could not provide a response to this survey question.

[80] https://www.urjc.es/universidad/calidad/2000-lengua-de-signos-espanola-y-comunidad-sorda#acceso-y-matriculacion

Table 4.6 Training on interpreting for deafblind persons		
Country / region	**Is interpreting for deafblind persons covered by any of the available formal training programs? If yes, at which training programs?**	
Austria	yes	*Universitätslehrgang Modus (Dolmetschen und Übersetzen für Österreichische Gebärdensprache, Deutsch und International Sign)* at the University of Salzburg
Denmark	yes	*Bachelor's degree in sign language interpreting* at Københavns Professionshøjskole
Finland	yes	*Tulkki (AMK), Viittomakielen ja puhevammaisten tulkkaus* at the Diaconia University of Applied Sciences (Diak)
France	yes	*Master Traduction, Interprétation et Médiation linguistique* at the Université Toulouse Jean Jaurès
Hungary	yes	*Sign language interpreting* at SINOSZ (2x)
Iceland	yes	*Táknmálsfræði og táknmálstúlkun* (Sign Language studies and interpretation) at University of Iceland
Italy	yes	*Various programs (5x)*
Netherlands	yes	*Bachelor Tolk Nederlandse Gebarentaal-NL*, HU University of Applied Sciences Utrecht
Norway	yes	*Norsk tegnspråk at OsloMet, Tegnspråk og tolking* at Høgskulen på Vestlandet (HVL), *Norsk tegnspråk* at NTNU
Poland	yes	*MA specialization in Polish Sign Language (PJM) within Translation Studies* at University of Warsaw, Institute of Applied Linguistics (ILS UW)
Portugal	yes	*Tradução e Interpretação de Língua Gestual Portuguesa* at Escola Superior de Educação do Instituto Politécnico de Setúbal, *Portuguese Sign Language Degree* at Escola Superior de Educação do Politécnico de Coimbra
Russia	yes	*Theory and practice of sign language interpreting in intercultural communication (RSL, English)* at Moscow State Linguistic University
Spain	yes	*Grado en Lengua de Signos Española y Comunidad Sorda* at Universidad Rey Juan Carlos
Sweden	yes	*Teckenspråkstolk- och dövblindtolkutbildningen* at Västanviks Folkhögskola, *Teckenspråkstolk- och dövblindtolkutbildning* at Fellingsbro Folkhögskola *Kandidatprogram i teckenspråk och tolkning, inriktning Tolkning mellan svenskt teckenspråk och svenska* at Stockholm University
Switzerland-Ge.	yes	*BA of Arts in Sign Language* Interpreting at Interkantonale Hochschule für Heilpädagogik

4.6 Interpreter training in specific topics

The students receive during their interpreter program training in specific topics, such as team interpreting and remote interpreting. As Figure 4.3 shows the topic of team interpreting is most prevalent and, so far, little attention is given to working with AI tools. A few countries mention the possibility for students to take other minors, but this is not common.

Fig. 4.3 Number of training program that educated students in specific topics

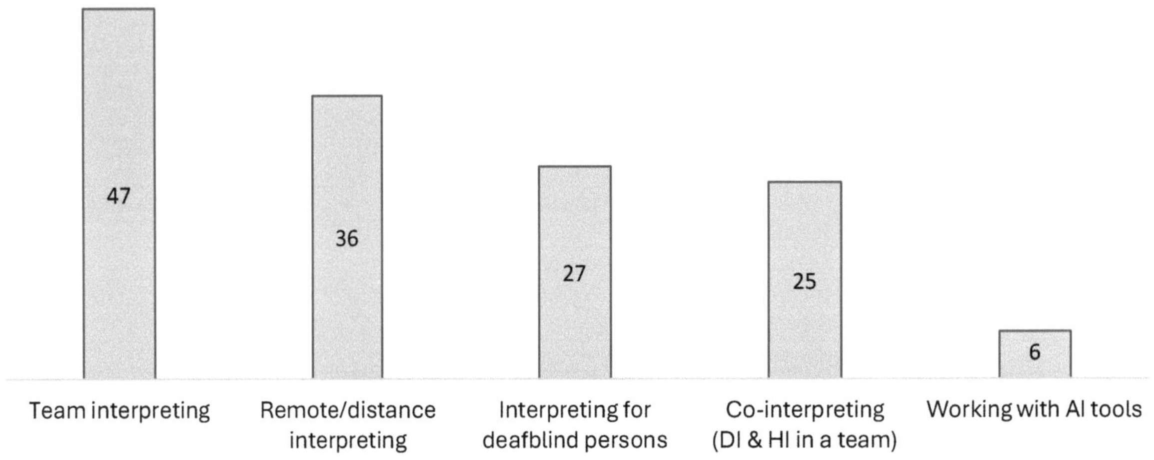

Fig. 4.4 shows that some programs teach additional languages. For example, in Belgium-Flanders you must choose French, English or German in the bachelor and in the master you can choose to only continue with Flemish Sign Language (VGT) or combine it with the other languages.

Fig. 4.4 Number of programs that teach additional languages

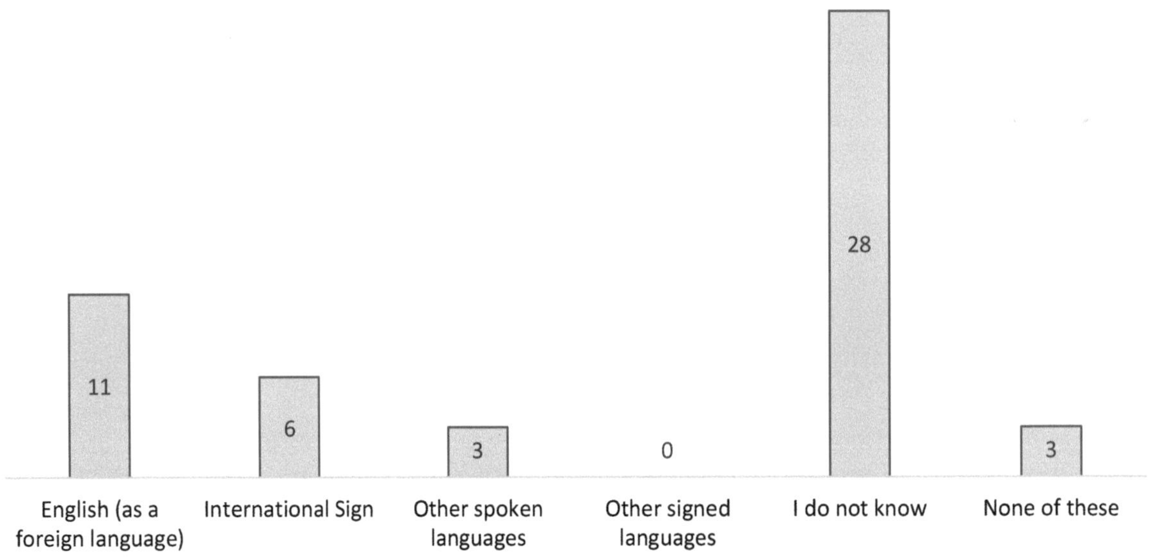

About three quarters of the training programs include an internship or a traineeship, whose duration varies from a few weeks to a full year (Fig. 4.5).

Fig. 4.5 Educational programs and if they have an internship or traineeship (n=68)

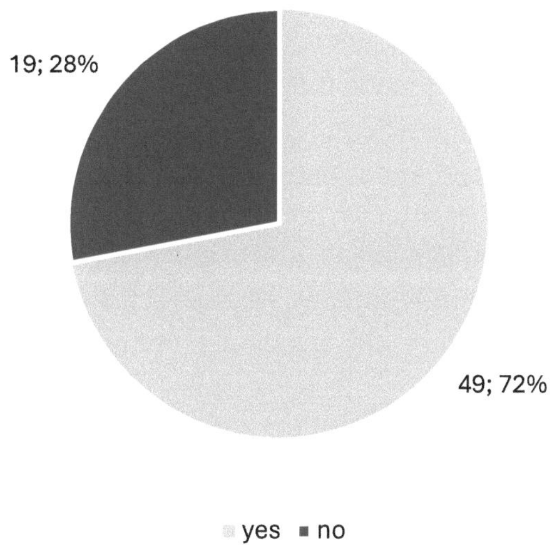

4.7 Interpreting degree as a prerequisite to work

In most of the countries an interpreting degree is required to work as a sign language interpreter (table 4.7). The requirement can also be related to the payment system in some of the countries. If the interpreter has an interpreting degree, they can be paid for their interpreting services by the responsible agency.

Table 4.7 An interpreting degree as a prerequisite to work as an interpreter

YES		NO
Austria	Latvia	Bulgaria
Belgium - Flanders	Norway	Czech Republic
Belgium - Wallonia	Portugal	Engl. W&NI
Denmark	Romania	Finland
Engl. W&NI – NUBSLI	Russia	France
Georgia	Scotland	Germany
Greece	Spain	Lithuania
Hungary	Switzerland– Ge.	Netherlands
Iceland	Switzerland– It.	Poland
Ireland		Sweden
Italy – ANIMU		Turkey
Italy - ANIOS		Ukraine[HI]

4.8 Continuing education

In most of the countries and the regions (65%) it is not required to undertake professional development to practice as an interpreter (Fig. 4.6). Some countries such as England, Wales & NI, Scotland and the Netherlands have an advanced system for continuing education and professional development for sign language interpreters. Their compulsory continued learning is directly linked to their national registration body, which monitors the quality of the training and tracks the professional development of each individual interpreter.

In Ireland, the Register of Irish Sign Language Interpreters (RISLI) did have CPD requirements, but these are not being followed at the moment. However, they are being reviewed and RISLI is seeking consultation with stakeholders with the view to it being reintroduce.[81]

There is no national accreditation or certification system for sign language interpreters in Poland. However, within the Association of Polish Sign Language Interpreters (STPJM), an internal system of professional recognition and continuing education is in place. STPJM members are required to complete a minimum number of Continuing Professional Development (CPD) hours each year. Interpreters who do not meet this requirement can maintain a passive membership status for up to three years. After that period, if they still do not fulfil the CPD criteria, their membership is terminated. This internal mechanism functions as the only structured form of maintaining professional standards, since no national framework for interpreter accreditation currently exists.

Malta mentions that there is no mandatory CPD but expect that this will be actualized through the new proposed law warranting sign language interpreters.[82] Slovakia and Belgium-Wallonia also mention the absence of mandatory CPD which they would like to see implemented in their country and region.

Table 4.8 lists all the countries and shows where continuing education is mandatory, how frequent and how many credits or hours are required.

---

[81] https://www.risli.ie/cpd-and-work-practice-policy-isl-summary-sept-2025/
[82] https://parlament.mt/14th-leg/acts/act-xxxviii-of-2025/

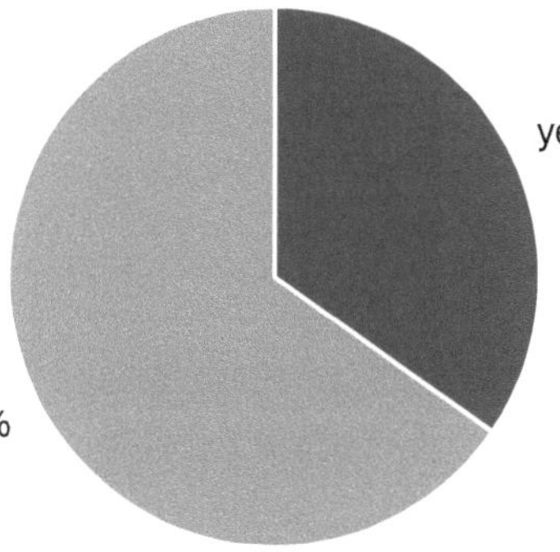

Table 4.8 Requirements of CPD / further training				
Country / region	**Mandatory**	**Frequency**	**Hours/Credits**	**Notes**
Albania	No	Not specified	Not specified	No formal requirements, occasional project-based training.
Austria	Partly	No fixed	Not specified	Required only for ÖGSDV members; most institutions do not require CPD after initial registration.
Czech Republic	Conditional	Yearly	24 hrs / year	Applies if working under Ministry of Social Affairs; otherwise depends on employer.
Engl. W&NI-ASLI	Yes	Yearly	24 points / year	Logged with NRCPD; structured points system.
Engl. W&NI-NUBSLI	Yes	Not specified	Not specified	CPD required as a condition for continued practice.
Estonia	Yes	Every 4 years	Not specified	Qualification renewal required; CPD needed for renewal.
Finland	Conditional	Not specified	Not specified	Required to remain in legal interpreter register (work activity and attend training).
Germany	Yes	Every 2 years	16 hrs / 2 years	At least 8 hrs must be interpreting-related; rest flexible.
Hungary	Yes	Every 3 years	60 hrs / 3 years	Structured requirement.
Italy-ANIMU	Yes	Yearly	Credits required	Annual training credits required to continue working.
Italy-ANIOS	Yes (members)	Yearly	Credits required	Needed to maintain legally required quality certification.
Kosovo	Informal	No specified	Not specified	General expectation to attend trainings; no formal system.
Latvia	Yes	Every 3 years	Not specified	Attend mandatory qualification improvement; certification needed for specific roles.
Lithuania	Yes	Every 3 years	36 hrs / 3 years	Structured requirement.
Netherlands	Yes	Every 4 years	Min. of 60 hours	Must collect minimum number of 60 hours per 4 years to maintain registration.
Switzerland-It.	Yes	Yearly	16 hrs / year	Plans for additional monthly CPD requirements.
Ukraine[HI]	Not specified	Not specified	Not specified	CPD through courses, seminars, workshops; no fixed system.

4.9 Interpreter specializations

In a total of twelve countries and regions it is possible to specialize in a certain topic or field (Fig. 4.7). Most countries and regions do not offer any specializations, due to different reasons:

- Too few interpreters in the country/region to offer or design a specialization: Iceland, Malta, Poland, Slovenia, Switzerland-French region
- Regional differences in the country, fewer interpreters less opportunities to specialize: France
- Specializations are acquired through professional and personal interest of the interpreter, and not through formal training: Ireland, Switzerland-French region, Engl.W&NI-NUBSLI
- Only ad hoc or one-time projects: Estonia
- Professional development is relatively new in the country/region: Russia, Turkey
- Interpreters work as all-round generalists: France, Poland, Slovenia

Russia mentions that they see a trend that in case of high remuneration in a certain area (e.g. conferences, media/TV), interpreters tend to specialize themselves but not through a specialized training.

Table 4.9 details the available and formalized specializations in the twelve countries and if they have any special requirements.

Fig. 4.7 Number of countries/regions with specializations (n=12)

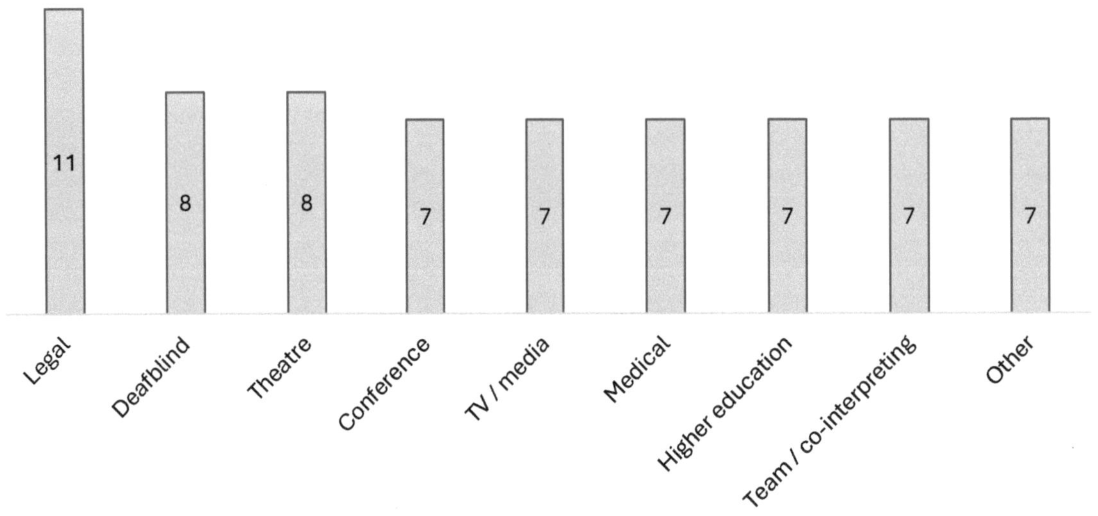

Table 4.9 Specializations available and/or formally recognized (n=12)

Country / region	Deafblind	Legal	Conference	Theatre	TV / media	Medical	Higher education	Team / co-interpreting	Other	Requirements for specializations	
	A = Available, F = Formally recognized										
Belgium – Flanders	A	A, F	A	A	A	A	A	A	A	no	
Czech Republic		F								yes	Legal interpretation (court, police...) has specific requirements: Like spoken language interpreters, modify for SL (no CzSL exam needed – we don't have it).
Denmark	A	A	F	A, F	F	F	F	F	F	yes	You must have an interpreting degree.
Engl. W&NI - ASLI	A, F			A							Specialisms are not generally attained via formal courses.
Engl. W&NI - NUBSLI	A	A	A	A	A	A	A	A	A	yes	Theatre Interpreting- main company is Performance Interpreting for big events. Strict regulation for medical interpreters and those in legal settings.
Finland	A, F	A, F								yes	Deafblind interpreting is a part of the bachelor's degree.
Germany	A	A, F	A, F	A	A	A	A	A	A	yes	The certification of an interpreter for the court is a formal swearing-in ceremony in which the interpreter takes an oath to render the content to be interpreted faithfully and conscientiously. To become certified, interpreters must apply to the relevant authority and demonstrate their personal suitability and professional expertise. This information is stored in a court database so that interpreters can be called upon when needed.
Italy - ANIOS	A, F	A	A	A	A	A	A	A	A		
Portugal		F									We do not have specializations in different areas; we only offer training in our association for professional improvement of those interested. It´s necessary to invest in different contexts.
Scotland	A	A	A	A	A	A	A	A	A	yes	Annual CPD.
Sweden		F				F					There is general authorization, authorization for work in legal settings and authorization for healthcare work at the agency ('Legal, Financial and Administrative Services Agency') Kammarkollegiet. For the general and healthcare authorization there is no training to attend, just self-studies.
Ukraine[HI]		A	A	A	A		A	A	A	no	Some interpreters get to know specializations on their own through practical experience, but I must point out that there aren't any training programs for obtaining specializations.

4.10 Mentoring

Nineteen countries and regions offer a form of mentorship or supervision (table 4.10). In 2020 only thirteen countries/regions reported the availability of a mentorship. The mentorships are mainly on a volunteer basis and seven of them are formally recognized programs (table 4.11). Only in six countries/regions the mentorship or supervision programs are mandatory. Ownership or the mentorship programs varies; there is a mixture of interpreter training programs (for interpreter students or trainees), interpreters themselves, interpreter agencies, deaf and deafblind associations or other organizations such as sign language interpreter associations (table 4.11).

Table 4.10 Availability of mentorship program for (student/trainee) sign language interpreters (n=49)

YES		NO	
Belgium - Flanders	Engl. W&NI - ASLI	Albania	Kosovo
Denmark	Engl. W&NI - NUBSLI	Austria	Moldova
Finland	Malta	Belgium - Wallonia	Montenegro
Georgia	Netherlands	Bosnia-Herzegovina	Romania
Greece	Norway	Bulgaria	Russia
Iceland	Poland	Croatia	Scotland
Italy-ANIOS	Portugal	Cyprus	Serbia
Latvia	Spain	Czech Republic	Slovenia
Lithuania	Switzerland – It.	Estonia	Sweden
Luxembourg		France	Switzerland – Fr.
		Germany	Switzerland – Ge.
		Hungary	Turkey
		Ireland	Ukraine[HI]
		Italy-ANIMU	

Table 4.11 Available mentorship program or supervision for (student/trainee) sign language interpreters

Country/region	Form	Type	Organized by
	M = Mandatory F = Formally recognized V = On a volunteer basis	S = Students PP = Peer to peer SP = Specific settings R = Recent graduates	TR = interpreter training program INT = interpreters themselves AGE = Interpreter agencies DB = deaf/deafblind association OTH = other
Belgium – Flanders	V	S, PP, SP	INT OTH: Tenuto, responsible for organizing CPD in Flanders
Denmark	V, F	PP	INT
Engl. W&NI - ASLI	V, F	S, R, PP, SP OTH: Professional supervision and mentoring are available for all career stages	INT, TR

Table 4.11 Available mentorship program or supervision for (student/trainee) sign language interpreters			
Country/region	**Form**	**Type**	**Organized by**
	M = Mandatory F = Formally recognized V = On a volunteer basis	S = Students PP = Peer to peer SP = Specific settings R = Recent graduates	TR = interpreter training program INT = interpreters themselves AGE = Interpreter agencies DB = deaf/deafblind association OTH = other
Engl. W&NI - NUBSLI	F, M	S, R, PP, SP	INT, TR, AGE, DB OTH: Specific to each program
Finland	V	PP	AGE OTH: interpreters' association, but not every year
Georgia	M	*S, R*	TR, DB
Greece	V	S, R, PP, SP	AGE
Iceland	F	S, PP	TR, AGE
Italy - ANIOS	V	R, PP	INT OTH: ANIOS has a mentoring program independently
Latvia	V	R	OTH: workplace where SLI work
Lithuania	F	S, R	AGE
Luxembourg	V	S	OTH: universities where students are enrolled
Malta	V	R, PP, OTH: for those employed indefinitely by the National Disability Agency	INT, AGE OTH: interpreter association
Netherlands	M	S	TR
Norway	V	S	TR
Poland	Not applicable	S, R, PP	OTH: SPTJM, the interpreter assoc.
Portugal	M	S	TR
Spain	F, M	S	TR
Switzerland – It.	F, M	S, R	TR, AGE

In Poland the first formal mentoring and supervision training for sign language interpreters was completed within the Association of Polish Sign Language Interpreters (STPJM) in October 2025. Starting in 2026, the newly trained mentors, both deaf and hearing, will begin conducting mentoring practice with PJM interpreters and interpreter trainees. From then on, mentoring and supervision services will be available both within STPJM's internal professional development system and on the open market, marking the first structured mentoring framework for sign language interpreters in Poland.

In Portugal and Spain, the only form of mentoring is during the interpreter training programs when students do an internship supervised by a senior interpreter who evaluates and monitors them. There is no mentoring program available after graduation.

Part 5 – Employment

5.1 Interpreters currently working

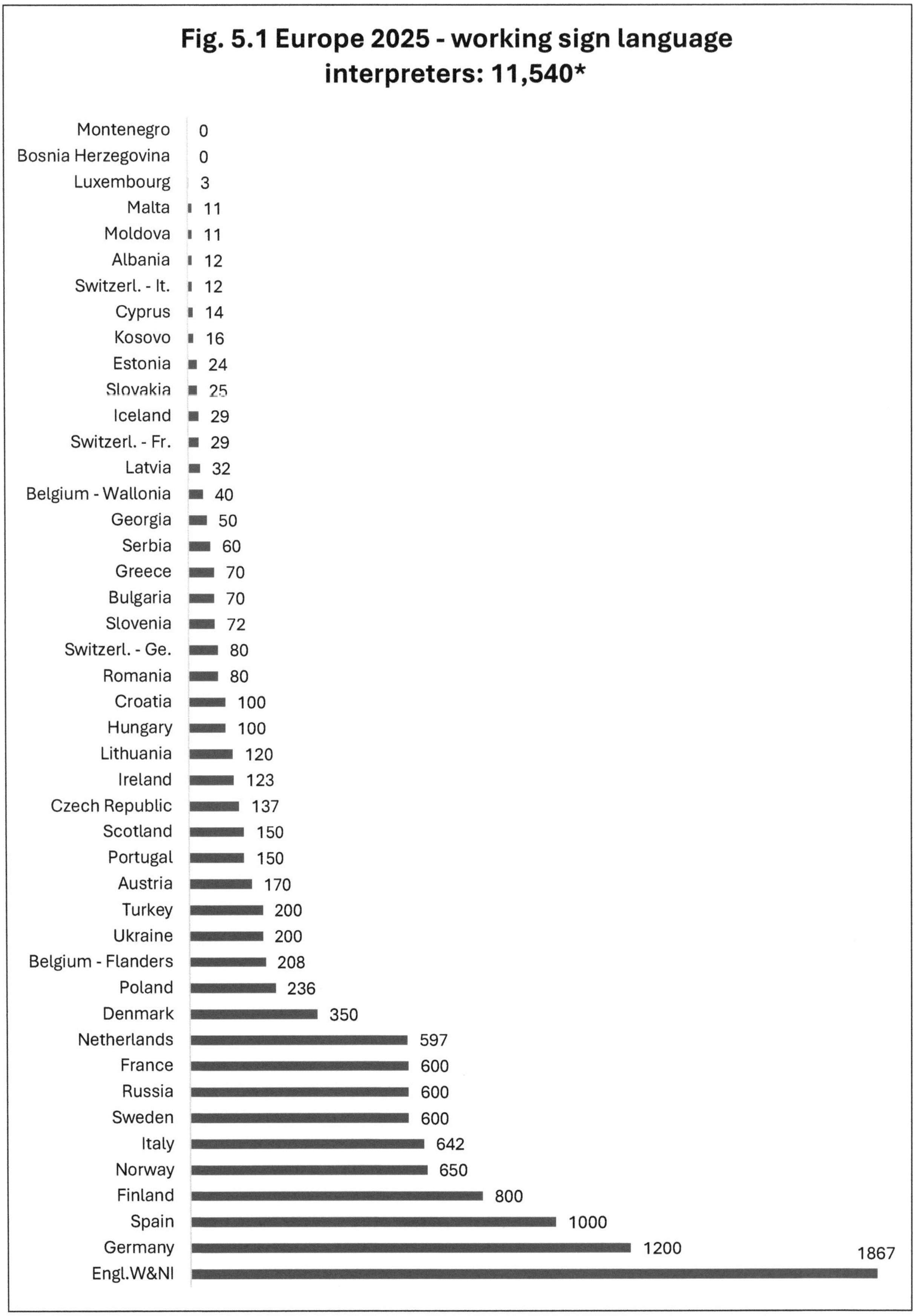

*Where a range was provided, the lowest number has been applied in the figure.

The estimated total number of sign language interpreters working in 2025 is 11,540 (Fig. 5.1). In 2020 the total estimated number of working interpreters was 10,083, which is a total increase of 14.5%.

Most of the respondents gave an estimate. In case the respondent provided a range of numbers, the lowest number was used in Figure 5.1. For Italy, the total number provided by ANIOS was taken (243) and the number of members of ANIMU (399) was added, as neither organization provided an estimate for all of Italy. For England, Wales & Northern Ireland (Engl. W&NI) the total number is taken from the NCRPD[83], the national registration body, and includes the registered sign language interpreters and the regulated trainee sign language interpreters.

Most countries struggle to provide an exact number of working sign language interpreters because of the absence of a national administration system. In addition, many countries do not have a registration body for sign language interpreters (see Part 2) and there is no requirement to have completed a training as a sign language interpreter to work (see 4.5). Inactive interpreters can, for example, still be registered. Countries with fewer interpreters are typically able to provide more precise numbers.

Some countries provided further explanations:

- Poland: according to the 2023 professional survey of sign language interpreters by Stowarzyszenie Tłumaczy Polskiego Języka Migowego (STPJM), there were 236 interpreters who responded to the census. This number reflects respondents to that survey and may not capture all individuals working as interpreters across the country.
- Turkey: The estimate of 200 sign language interpreters includes not only those who have passed the sign language interpreting exam, but also individuals who have completed the master's program in sign language interpreting. In addition, it includes interpreters working in the activities of various non-governmental organizations, those employed in TV shows and news programs, as well as Children of Deaf Adults (CODAs) who are active in this field.
- Ukraine[HI]: This data does not include the more than forty interpreters who were forced to leave Ukraine, particularly to EU countries, searching for refuge after Russia's full-scale invasion of Ukraine in early 2022. Nor does it include those who were left in the Russian-occupied territory of Ukraine.

The twenty-seven member states of the European Union currently have an estimated number of 7,574 working sign language interpreters (Fig. 5.2). That is a total increase of 836 interpreters since 2020. Table 5.1 provides an overview of the total number of sign language interpreters in the EU in 2016, 2020, and 2025. In most countries there is an increase in the number of interpreters in the past ten years, except in the following six countries: Cyprus, Estonia, Finland, Hungary, Latvia, and Portugal.

[83] https://www.nrcpd.org.uk/registration-figures

Fig. 5.2 EU member states 2020 > 2025
working sign language interpreters: 6,738 > 7,574

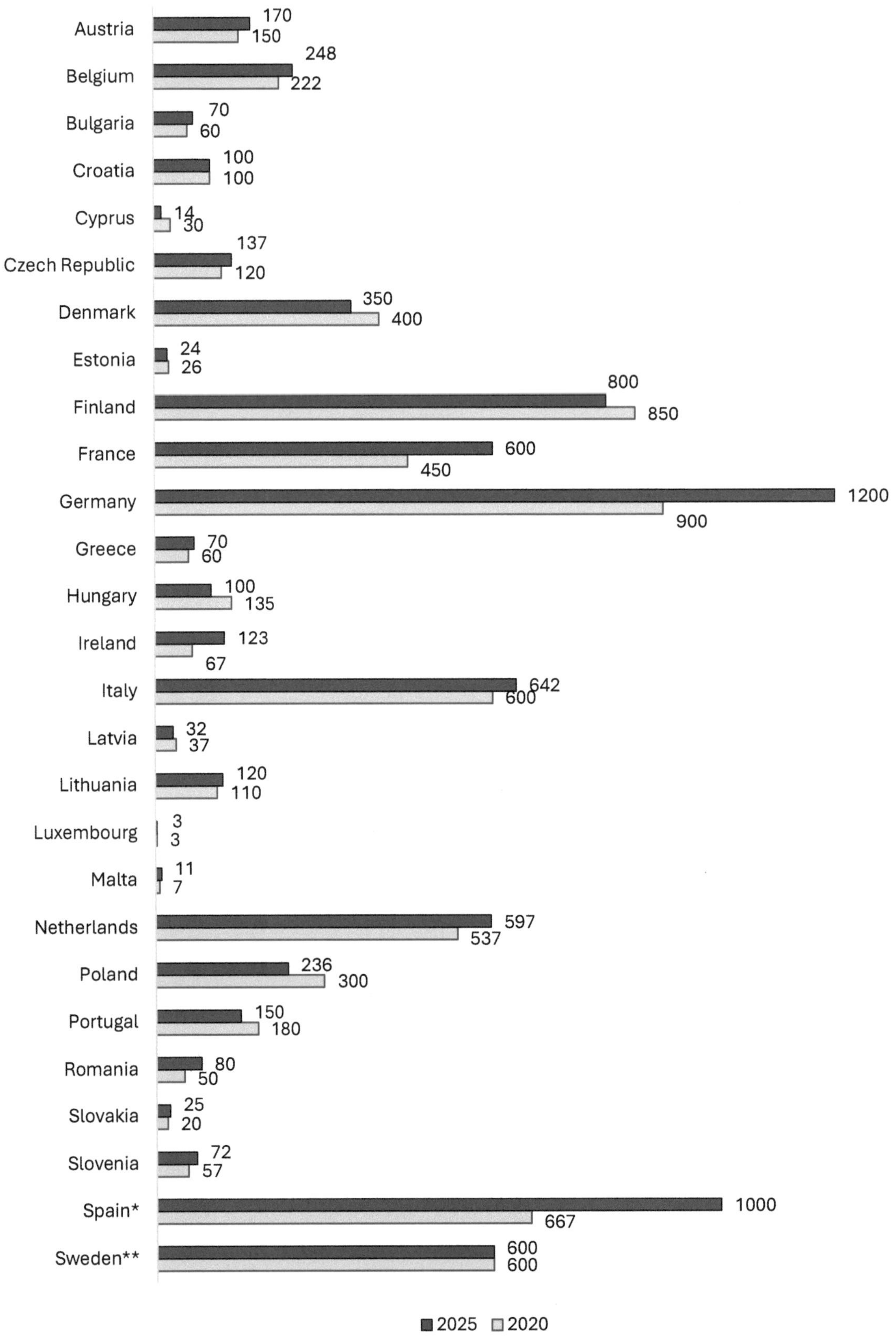

** The 2020 total number is taken from the 2012 survey*

*** The 2020 total number is taken from the 2016 survey*

Table 5.1 EU member states 2025, 2020, 2016 - working sign language interpreters					
Country	**2025**	**2020**	**2016**	**Change in number of members 2016>2025**	**%**
Austria	170	150	150	+20	12
Belgium	248	222	189	+26	24
Bulgaria	70	60	46	+10	34
Croatia	100	100	60	0	40
Cyprus	14	30	25	-16	-79
Czech Republic	137	120	70	+17	49
Denmark[84]	350	400	400	-50	-14
Estonia	24	26	25	-2	-4
Finland	800	850	600	-50	25
France	600	450	400	+150	33
Germany	1200	900	750	+300	38
Greece	70	60	50	+10	29
Hungary	100	135	150	-35	-50
Ireland	123	67	50	+56	59
Italy	642[85]	600	200	+42	69
Latvia	32	37	41	-5	-28
Lithuania	120	110	111	+10	8
Luxembourg	3	3	2	0	33
Malta	11	7	8	+4	27
Netherlands	597	537	499	+60	16
Poland	236	300	200	-64	15
Portugal	150	180	100	-30	33
Romania	80	50	12	+30	85
Slovakia	25	20	35	+5	-40
Slovenia	72	57	46	+15	36
Spain	1000	667[86]	667	+333	33
Sweden	600	600*	600	0	12
TOTAL	**7,574**	**6,738**	**5,486**		

Note: the UK left the EU on 31 January 2020 and is not included in table 5.1

[84] The total number is taken from the 2016 survey
[85] The total number is the number provided by Italy ANIOS + total number of Italy ANIMU interpreter members
[86] The total number is taken from the 2012 and 2016 survey

5.2 Part-time versus full-time

Figure 5.3 shows the percentage of the sign language interpreters that work full-time versus part-time. In only includes the countries that were able to give a number. However, most respondents indicate that the percentage is an estimate. France, Ireland, Netherlands, Norway were able to give a more precise number based on national surveys they conducted during the past years.

Fig. 5.3 Interpreters working full-time vs. part-time (%)

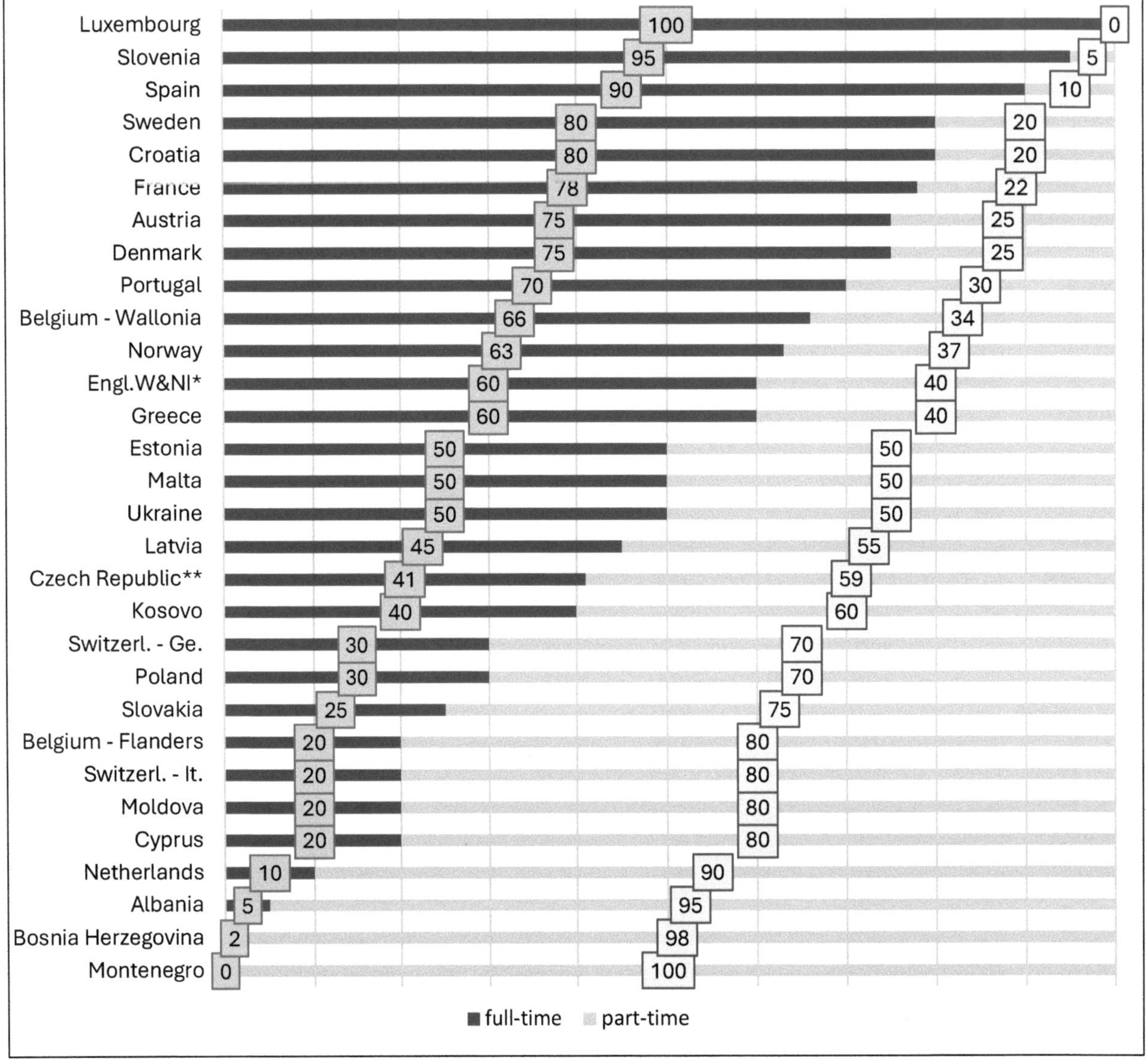

**Based on the 2021 UK census*[87]

***Based on the 2024 survey in Czech Republic*[88]

****Albania has five full-time interpreters: two at the Deaf Resource Center, one at the University, and two at ANAD.*

[87] https://asli.org.uk/files/download/48511768939db4f

[88] https://www.cktzj.com/scitani-tlumocniku/

5.3 Interpreting for deafblind persons

Continuing from the 2020 survey, the respondents answered how many sign language interpreters in their country or region interpret for deafblind persons. The countries that were able to provide a number or an estimate are presented in Figure 5.4. Countries and regions that do not have this data are Belgium-Flanders, Belgium-Wallonia, France, Germany, Greece, Hungary, Ireland, Italy-ANIMU, Norway, Russia, and Scotland.

Figure 5.4 shows that the data varies greatly per year (2020 and 2025) and that there is no robust system of registration concerning interpreting for deafblind persons.

Turkey mentions that there are officially no interpreters that work specifically for deafblind persons or are as trained as such. In Estonia and Sweden all interpreters are nominally qualified to work for persons who are deafblind. Slovakia mentions the same, but states there are very few known deafblind persons that request sign language interpreting services, about eight. As a result, the interpreting is done by the same ten interpreters that the deafblind persons know. Latvia mentions that there are no requests for interpreting for deafblind persons.

Poland estimates based on the available data and past certification records, that there are around 100 interpreters in Poland who are trained to work with deafblind persons. This estimate comes from the number of certificates issued by the Society for the Care of the DeafBlind (Towarzystwo Pomocy Głuchoniewidomym - TPG), which operated until 2019. Since the organization no longer exists and no new certification body has replaced it, the current number is difficult to verify, and the actual number of active deafblind interpreters may be lower.

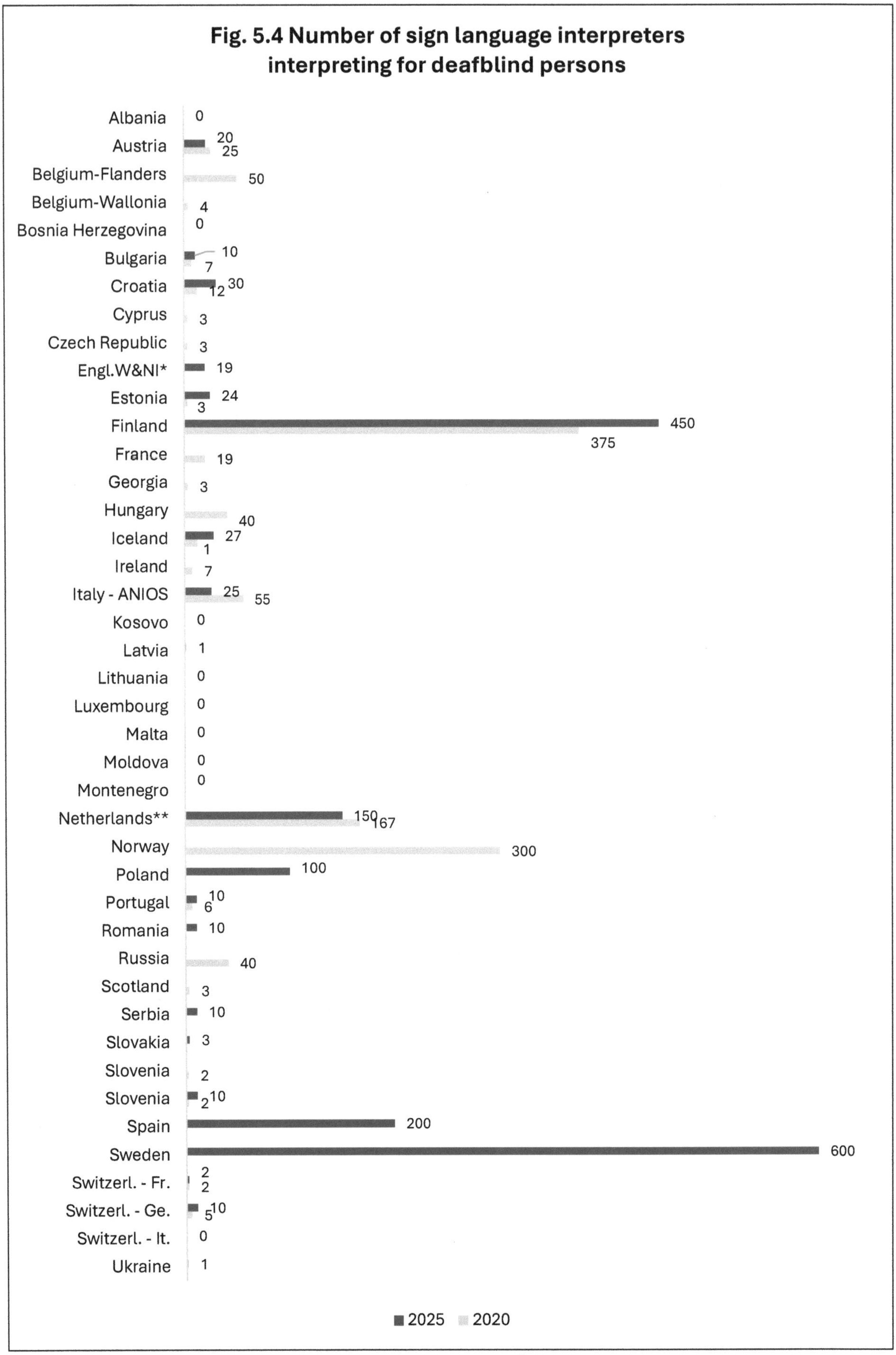

*According to NCRPD (https://www.nrcpd.org.uk/registration-figures)

**According to the national registry (stichtingrtgs.nl)

5.4 Lack of interpreters

Throughout Europe 71% of the respondents report a shortage of interpreters (Fig. 5.5). There are several reasons why the demand cannot be met:

- Ratio: Very high demand, too few interpreters.
- Varied high and low demands, depending on:
  - Days of the week
  - Regions
  - Settings, e.g. education
  - Additional competencies.
- Training:
  - Lack of education
  - Education is self-funded
- Leaving the profession:
  - Retirement
  - Job insecurity
  - Inadequate working conditions
  - Absence of professional recognition.
- Funding: not enough funding to pay for interpreting services.
- War in Ukraine: Interpreters forced to flee the country.

Greece, Latvia, Serbia, Spain, Sweden and Switzerland-Italian region state that there are enough interpreters to meet the demand. Spain explicates the difficulties in the profession:

> "The main problem lies in the job insecurity that plagues this profession, causing many interpreters to leave the sector and pursue other, better-paid professions with greater social recognition. Pressure from contracting entities, coupled with poor working conditions, contributes to this situation. Salaries are extremely low, contracts are unstable, working hours are excessive and unpredictable, and a lack of respect and professional recognition is commonplace. All of this creates a climate of demotivation and brain drain, directly impacting the quality of service and the future of the profession."

Fig. 5.5 Sufficient number of interpreters to meet the demand?

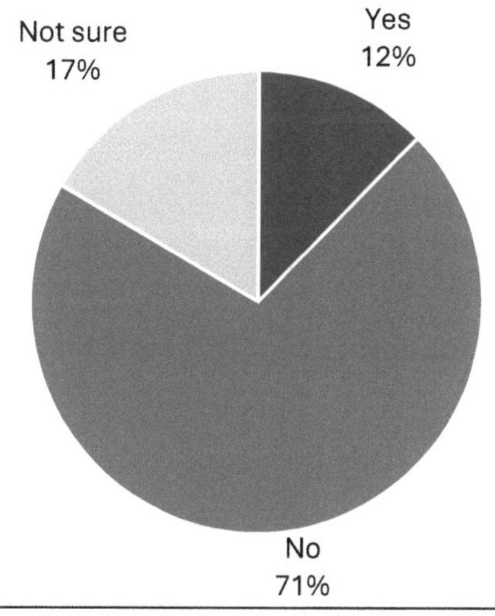

In Albania there has been a shortage of interpreters for years. However, a project by the Finnish Deaf Association and UNDP Albania trained and certified 7 new interpreters (five deaf and two hearing interpreters), bringing the total to 12 interpreters in Albania in 2025.

Fig. 5.6 shows where the lack of interpreters is deemed to be the highest: in remote regions and for last-minute assignments.

Fig. 5.6 If there is a lack of interpreters, where is this lack noticeable (n=37)

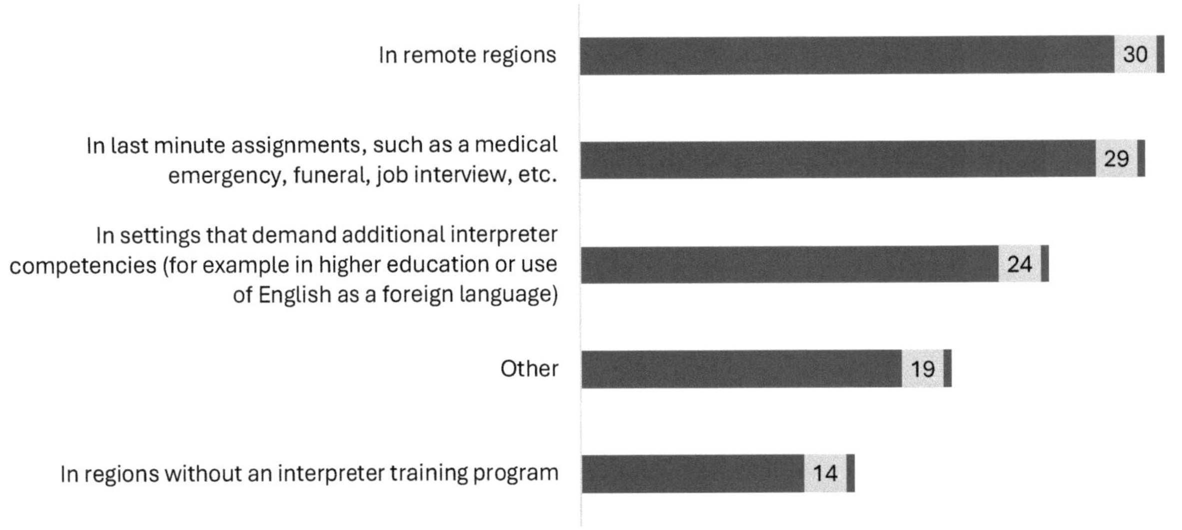

5.5 Number of working interpreters: 2001 - 2025

From the very first report in 2001 onward, there have been fifteen respondents who participated in each of the six surveys (2001, 2004, 2007, 2012, 2016, 2020, 2025): Austria, Belgium-Flanders, Denmark, England, W & NI, Estonia, Finland, France, Ireland, Italy, Netherlands, Norway, Portugal, Scotland, Switzerland-French region, and Switzerland-German region.

Table 5.2 shows the development in these fifteen countries during this period (2001-2025). In 2001 there were reportedly 1,778 interpreters and in 2025 the total is 6,440 interpreters. All fifteen countries showed an increase in the number of working interpreters.

Table 5.2 Total number of sign language interpreters in fifteen countries							
Year	**2025**	**2020**	**2016**	**2012**	**2007**	**2004**	**2001**
Total	**6,440**	**5,187**	**4,253**	**3,787**	**3,016**	**2,277**	**1,778**
Austria	170	150	150	120	100	100	100
Belgium-Flanders	208	185	165	160	150	136	100
Denmark	350	400[89]	400	300	300	200	100
England, Wales & NI	1,867	1,000[90]	1,000	700	450	350	350
Estonia	24	26	25	25	27	27	25
Finland	800	850	600	600	450	200	255
France	600	450	400	360	180	150	100
Ireland	123	67	50	60	40	20	25
Italy	642[91]	600	200	300	300	300	250
Netherlands	597	537	499	400	207	159	65
Norway	650	550	500	500	550	500	200
Portugal	150	180	100	100	137	15	24
Scotland	150	90	69	80	53	53	32
Switzerland-Fr.	29	32	32	30	30	25	17
Switzerland-Ger.	80	70	63	52	42	42	35

5.6 Number of deaf signers

Figure 5.7 shows an estimate of the number of deaf signers in forty-five countries and regions. It is difficult for the respondents to provide precise data, but they give estimates based on, for example, services or deaf persons enrolled in education. Many countries lack national census data on the number of deaf signers amongst their population. Some countries have statistics on persons with disabilities, but do not provide aggregated data on deaf, hard of hearing, deafblind, late deafened persons and their preferred modes of communication.

The number of deaf signers has also changed since the 2020 publication. For some countries more precise data is now available or new estimates have been made. Especially countries with larger populations find it difficult to give an estimate. The European Union of the Deaf published new estimates in their 2025 publication,[92] which have been checked by the national associations of the deaf. That said, the definition of 'deaf signers' may vary per country and altogether robust data collection methods have not yet been implemented.

[89] The total number is taken from the 2016 survey
[90] The total number is taken from the 2016 survey
[91] The total number is the number provided by Italy ANIOS + total number of Italy ANIMU interpreter members
[92] https://eud.eu/from-recognition-to-officialisation-a-new-eud-book-redefining-sign-language-rights-in-europe/

For Spain the number varies greatly, according to the national statistical office of Spain (la Instituto Nacional de Estadística (INE)) there are 27,300 persons over the age of 6 that use Spanish Sign Language (LSE). According to CNSE (National Federation of deaf persons in Spain) the number is about 100,000 persons.

Furthermore, the interpreter from Ukraine explains that Russia's 2022 invasion and the ensuing war make it impossible to provide even approximate information about the number of deaf signers in Ukraine.

In summary, the estimated number of deaf signers is more precise if there is a related question in the national census, such as in Scotland and England, W&NI. It also seems to depend on the use of services by the deaf signers. If a deaf person is requesting for example interpreting services through a formal system, then they are formally documented as deaf signers. This means that when there is a well-functioning system of interpreting services, it will be easier to identify the number of deaf signers. However, those that do not use any sign language interpreting services might not be officially documented anywhere as a deaf signer. In addition, it seems that how deaf persons are educated (mainstreamed with an interpreter or at a deaf school), has an impact on the number of signers overall.

In conclusion, speculating, the number of deaf signers per country seems to be anywhere between 1 in 600 to 1000 inhabitants.

5.7 Deaf signers per interpreter

In the forty-five countries and regions, there are more than 1,2 million deaf signers (Fig. 5.7) and over eleven thousand sign language interpreters in Europe. Figure 5.9 gives an average per country of the number of deaf signers (1,221,820) and working sign language interpreters (11,540). The average number of deaf signers per interpreter is 106. For the EU member states the average is 90.

Fig. 5.7 Deaf signers (45 countries & regions): 1,221,820

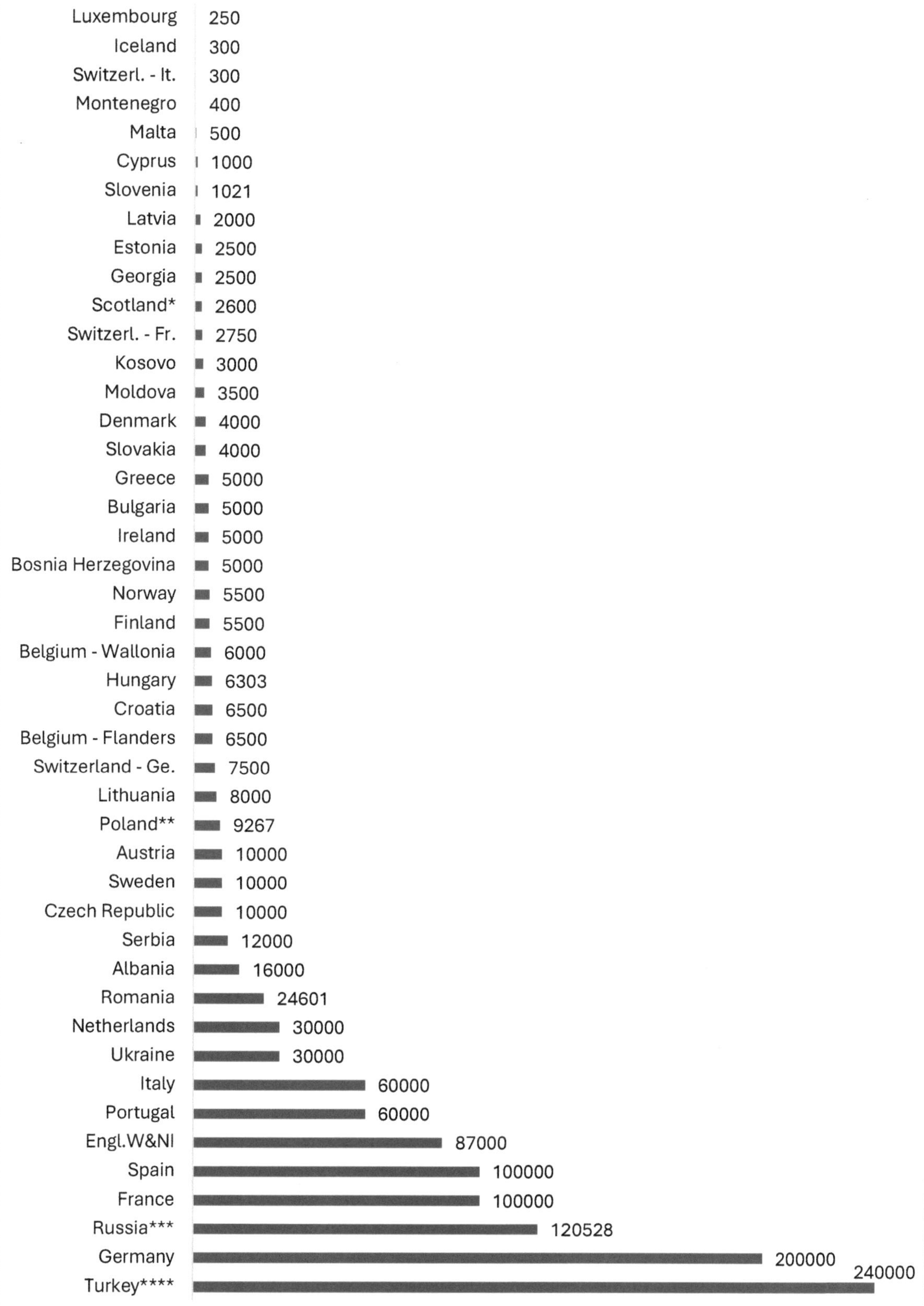

**Scotland according to the 2022 National Census93*

***Poland according to the 2021 National Census94*

****Russia, based on https://minlang.iling-ran.ru/en/lang/russian-sign-language*

*****Turkey: ~240,000-390,000 TİD users, based on: https://thesignhub.eu/node/1142*

Fig. 5.8 Deaf signers in the EU: 682,942

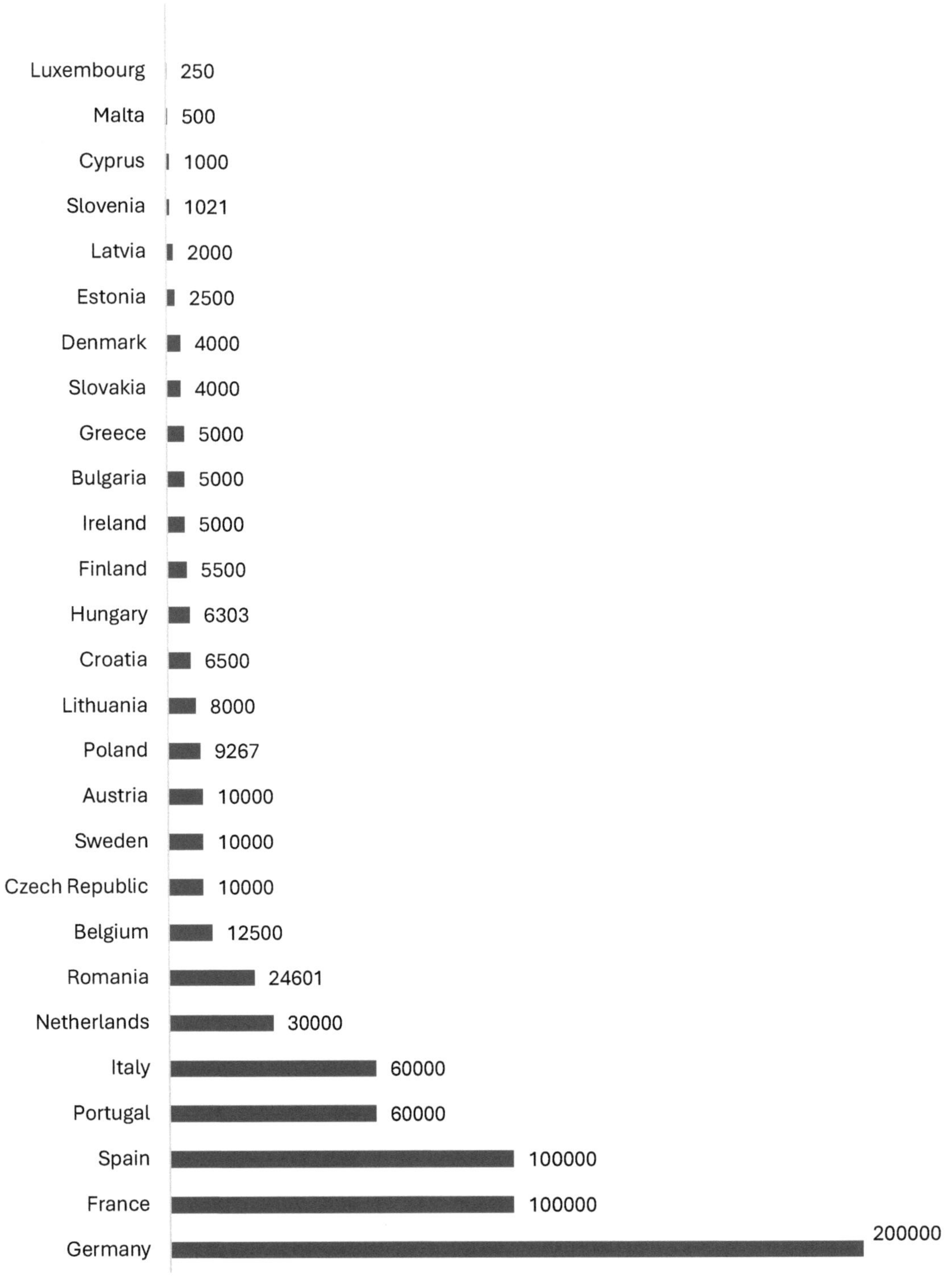

[93] https://www.scotlandscensus.gov.uk/2022-reports/scotland-s-census-2022-ethnic-group-national-identity-language-and-religion/#section8

[94] Published by the Central Statistical Office (GUS) in October 2023: 9,267 people in Poland declared Polish Sign Language (PJM) as the language they use at home. This is the first official and data-based estimate of the signing population in Poland. It includes Deaf and hard of hearing persons who identify PJM as their primary or home language. The actual number of PJM users—when including CODAs and other signers who use PJM as a second language—is likely higher.

Fig. 5.9 Deaf signers per interpreter (43 countries & regions)*

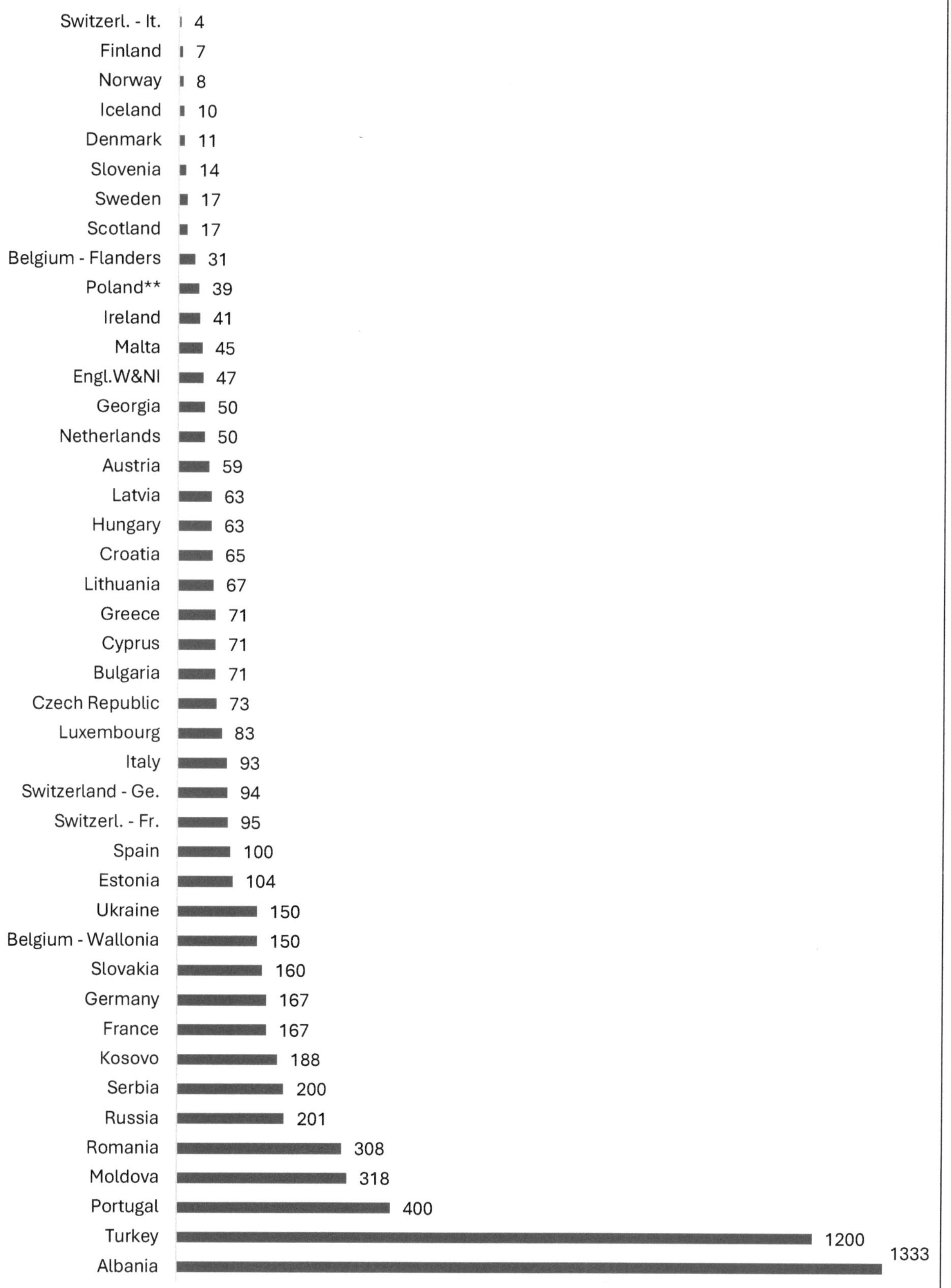

**Not including Bosnia-Herzegovina, Montenegro*

***Poland according to the 2021 National Census*

5.8 The right to an interpreter

Each country in Europe has different rights, regulations, and laws in regard to sign language interpreting. The first EU Directive on the Rights to Interpretation and Translation in criminal proceedings was adopted in October 2010 by the EU Council of Ministers[95]. This directive only gives the right to an interpreter or translator in criminal proceedings and not in other settings.

In 2017 the European Deaf Students' Union published a report 'Handbook for Students in Europe[96]. The report focuses on the legal issues that deaf students often face when studying abroad. It also shows that countries do not recognize the right to an interpreter for those studying abroad.

The countries and regions were asked in the survey if there is a law or regulation which gives the right to a sign language interpreter. These answers are shown in table 5.3. There are six countries where deaf persons do not have a right to a sign language interpreter: Albania, Belgium-Wallonia, Bosnia-Herzegovina, Cyprus, Greece, and Montenegro.

As table 5.3 shows, there is an enormous diversity in the right to an interpreter in the countries and regions. A country may entitle deaf signers to interpreting services without having sufficient interpreters to fill the demand.

Table 5.4 shows from what age onward a deaf or deafblind child or person has the right to an interpreter in an educational setting for those countries or regions that were able to provide an answer.

95. http://www.consilium.europa.eu/uedocs/cms_data/docs/pressdata/en/jha/116913.pdf
96 https://edsunion.wixsite.com/edsu/handbook

Table 5.3 Right to a sign language interpreter	
Country / region	**What do these rights consist of?**
Right to a sign language interpreter or interpreting services recognized by country/region laws or regulations	
Belgium - Flanders	Deaf people receive a package of interpreting hours from the government in different domains (Education, Work and Private Life).
Bulgaria	From the BGSL Act Art. 20. (1) Deaf and deaf-blind persons are entitled to free translation services in Bulgarian sign language up to an annual limit of 120 hours. (2) Deaf and deaf-blind students and doctoral students in the process of training for third level education, in addition to the limit provided under para. 1, are entitled to additional use of interpreting services in Bulgarian sign language up to 60 hours per semester. (3) The number of hours for free interpreting services in Bulgarian sign language is determined for each calendar year in accordance with the limit under para. 1 and 2, in proportion to the term of validity of the expert decision of TELK or NELK or the remaining months after the provision of the interpreting service. (4) Deaf and deaf-blind persons are entitled to free interpreting services beyond the limit provided under para. 1 and 2 in Bulgarian sign language in the following: 1. hospitalization in medical institutions for inpatient care, mental health centers, centers for skin and venereal diseases and complex oncology centers; 2. (amended - SG, issue 10 of 2023, effective 31.01.2023) performance by the competent authorities of investigative, judicial and other procedural actions under the Criminal Procedure Code, as well as when exercising the powers under the Ministry of Interior Act.
Croatia	Deaf, hard of hearing, and deafblind individuals can access interpreting services based on the number of hours per month granted by the government. Typically, hard of hearing persons receive up to 20 hours per month, deaf individuals up to 40 hours, and deafblind users around 160 hours.
Czech Republic	Czechia signed also EU-acts and norms.
Denmark[DI]	The right to a sign language interpreter is recognized in Danish law. The Act on Interpreting for Persons with Hearing Disabilities (2013) guarantees access to interpreting services for deaf, hard-of-hearing, and deafblind persons. These rights consist of: Free interpreting services for private life, leisure, health care, and contact with public authorities. Unlimited interpreting ("tidsubegrænset tolkning") in essential areas such as medical appointments and communication with authorities. Grant-based interpreting ("bevillingstolkning") for other activities not covered by unlimited interpreting (e.g. leisure activities, family events). The interpreting is administered by the National Interpreting Authority (DNTM), which assigns services through registered providers. Education and employment are not included in this law but are covered by other ministries and schemes.
Engl. W&NI	Equality act 2010 places duty for 'reasonable adjustments' for disabled people including right to interpreter for deaf people to access services.
Estonia	The right to use Estonian Sign Language in official communication with the state and local governments. The right to have the interpreting services provided by the local governments in everyday situation. And the right of learning in Estonian Sign Language.
Finland	A right to have interpreting service so that a person can lead a normal life like any hearing person would (work, study, have hobbies, take part in the society etc.) on the condition that they are able to express themselves and have means to communicate in a way or another.
France	Every public service must be accessible to any disabled person. For Deaf and Deafblind, it means they have the right to interpreters, and the cost will be taken care of by the public entity. It is also said that education must be given in the language chosen by the family. But the number of bilingual schools is not sufficient enough to respond to the demand.
Germany	People with hearing impairments (deaf, deafened, and hard-of-hearing people) and people with speech impairments have the right, in accordance with the relevant laws, to use German sign language, signs accompanying spoken language, or other suitable communication aids. At the request of the entitled persons, public authorities shall provide the appropriate communication aids free of charge or bear the necessary expenses.
Iceland	Rights of patients to understand treatment, in the justice system, in schools - other interpretation is not in the law.

Table 5.3 Right to a sign language interpreter

Country / region	What do these rights consist of?
Ireland	A public body shall do all that is reasonable to ensure that interpretation into Irish Sign Language is provided for a person who is competent in that language and cannot hear or understand English or Irish when that person is seeking to avail of or access statutory entitlements or services provided by or under statute by that public body. https://www.irishstatutebook.ie/eli/2017/act/40/enacted/en/print#sec1
Kosovo	Right to interpreting to the public sector
Latvia	Cabinet of Ministers Regulation No. 250 of 20 April 2021 "Procedures by which the Latvian Society of the Blind and the Latvian Association of the Deaf provide social rehabilitation services and provide technical aids – typhlotechnics and surdotechnics"; Cabinet of Ministers Regulation No. 338 of 13 June 2017 "Requirements for Social Service Providers".
Lithuania	Government law.
Luxembourg	Especially in educational settings, but also in public service settings. -Interpreter when contacting authorities, - Every hearing-impaired or deaf student in Luxembourg has the right to sign language instruction as well as the right to complete primary and secondary school in sign language. (4) Parents and siblings of a hearing-impaired or deaf person who uses this language as their predominant language are, if they are resident in the Grand Duchy of Luxembourg, entitled to basic training in sign language https://hoergeschaedigt.lu/online/www/vertical/information/4747/4748/containerMiddle/4752/1188/DEU/20170531_PDL_DGS_Commentaire%20articles-DE.pdf https://hoergeschaedigt.lu/online/www/horizontal/downloads/DEU/index.html Law on sign language recognition: https://men.public.lu/content/dam/men/fr/actualites/articles/communiques-conference-presse/2017/06/29-gebaerdesprooch/projet-de-loi.pdf
Malta	The National Disability Agency employs sign language interpreters to provide sign language interpreting services free of charge to Deaf sign language users.
Moldova	There is a legal framework that guarantees the right to communication through a sign language interpreter, both in social and administrative relations, and in justice.
Netherlands	https://wetten.overheid.nl/BWBR0049587/2024-04-19
Norway	Right to an interpreter, paid by the state (social services) for interpreting in daily life, work, education, health.
Poland	The right to a sign language interpreter is recognized in Polish law, primarily through the Accessibility Act (Ustawa o zapewnianiu dostępności osobom ze szczególnymi potrzebami, 2019) and earlier through the Act on Sign Language and Other Means of Communication (Ustawa o języku migowym i innych środkach komunikowania się, 2011). These laws guarantee that: Public institutions (including hospitals, offices, universities, and courts) must ensure access to communication for Deaf and hard of hearing persons, including through sign language interpreting services. Deaf persons have the right to choose their preferred mode of communication (PJM, Signed Polish, or written Polish). Each public body is required to provide accessibility upon request, usually within 14 working days. However, these rights are activated only when the person formally requests accessibility. If no official request is submitted, the institution is not obliged to provide an interpreter. In practice, this limits spontaneous or immediate access and places the administrative burden on the Deaf person. While the legal right exists, implementation is inconsistent, and the quality or availability of interpreters depends largely on local resources rather than a centralized national system.
Portugal	It is recognized that it is a language and that an interpreter/translator is necessary. Without one, communication is difficult. In the educational context, in reference schools, in the justice system in situations involving deaf people. In other contexts, whenever there is a deaf person and no other means of communication is possible, there is already concern about finding solutions.
Romania	Deaf people have the right to an interpreter in any situation and the right to receive education in Romanian Sign Language.
Russia	Government-funded 84 hours of interpreting
Scotland	All individuals and communities in Scotland have the right to access and understand full, accurate and timely information, in a manner appropriate to their needs to be included in, to benefit on an equal basis from, all the opportunities and services offered in their local communities.

Table 5.3 Right to a sign language interpreter	
Country / region	**What do these rights consist of?**
Serbia	The rights to a sign language interpreter in Serbia consist of the ability for Deaf and hard-of-hearing individuals to access spoken information in sign language. This includes communication in public services, education, healthcare, and legal settings, such as courts and police. However, these rights are not fully guaranteed in practice, as there are limited numbers of accredited interpreters, and formal systems for training, certification, and professional development are lacking.
Slovakia	Under Act No. 149/1995 Coll. on the Sign Language of Deaf Persons, the right to an interpreter means that Deaf persons are entitled to communication access through a qualified sign language interpreter when interacting with hearing people, particularly in situations where spoken Slovak is used. This right ensures that Deaf persons can: communicate effectively in official settings (e.g. with public authorities, healthcare providers, or schools), access information and services provided in the state language, and participate equally in social, cultural, and public life. The law requires that interpreting be provided between Slovak Sign Language and the state language. However, it does not specify detailed procedures (for example, how to request an interpreter or who pays for the service); those aspects are often covered by other regulations or social service frameworks, such as Act No. 448/2008 Coll. on Social Services, which defines interpreting services as one of the forms of social assistance available to Deaf persons.
Slovenia	A deaf person has the right to use sign language in procedures before state bodies, local self-government bodies, public authority holders or public service providers. A deaf person also has the right to use sign language in all other life situations in which deafness would constitute an obstacle to meeting their needs. A deaf person has the right to be informed in techniques adapted to them in accordance with special regulations. The right from the first and second paragraphs of this article is exercised by exercising the right to a sign language interpreter. State bodies, local self-government bodies, public authority holders or public service providers are obliged to provide a deaf person with a sign language interpreter at the request of the deaf person or ex officio as soon as the deaf person presents for inspection a document recognizing their right to an interpreter. The right to an interpreter is exercised to the extent and in a manner that enables the exercise of rights under the principle of equal accessibility. Payment of the costs of a sign language interpreter referred to in the first paragraph of this article shall be ensured by state authorities, local self-government authorities, public authority providers and public service providers in their annual financial plans.
Spain	According to the law—though not in common practice—interpretation services should adhere to the principles of quality, accessibility, and professionalism. However, these services are primarily funded through public subsidies managed by representative organizations of deaf people, which administer these resources and organize the service according to their own criteria, without a uniform framework for external control or oversight, or by private entities that are also not subject to any kind of control.
Sweden	The right to an interpreter in all areas of live, provided by the government. But the systems are complicated and tricky to navigate and the responsibility to provide interpreters is very vague in the private sector.
Switzerl.- Fr.	The right to an interpreter is not explicitly recognized by law, but accessibility provisions under the Federal Disability Equality Act (LHand) and the Invalidity Insurance Act (LAI) allow for interpreting services in specific contexts such as work, education, and access to public services.
Switzerl.- Ge.	Switzerland has ratified the CRPD and we have the Federal Act on the Elimination of Discrimination against People with Disabilities.
Turkey	It was recognized through a regulation issued in 2006. https://www.mevzuat.gov.tr/mevzuat?MevzuatNo=10601&MevzuatTur=7&MevzuatTertip=5 https://www.mevzuat.gov.tr/mevzuat?MevzuatNo=10123&MevzuatTur=7&MevzuatTertip=5
Ukraine[HI]	The law regulates the use of sign language interpreters in legal and social settings.

Table 5.4 From what age onward does a deaf or deafblind person have the right to an interpreter in an educational setting?	
Country / region	**Age / Condition for Right to Interpreter**
Belgium-Flanders	From age 2.5
Bulgaria	No age limit
Croatia	From kindergarten
Denmark[DI]	From age 0
Engl. W&NI	Varies; often limited access
Estonia	Responsibility on school, not age-based right
Finland	When child can communicate (no fixed age)
France	From birth (0), preference for signing professionals
Georgia	Any age
Germany	Possibly from birth
Iceland	From birth (0)
Ireland	Case-by-case from primary/secondary school
Italy	From first year of school, at all education levels
Latvia	From age 16
Lithuania	From kindergarten if child knows sign language
Luxembourg	No fixed age (depends on parents)
Malta	From age 3 (or earlier in childcare)
Moldova	No rights/guarantees in education
Montenegro	No rights/guarantees
Netherlands	From entry into education
Norway	No fixed age (depends on location/school)
Poland	Possible from entry into education (including preschool)
Portugal	Around 5th grade (earlier possible in reference schools)
Romania	Not specified
Russia	Based on hours, not age
Serbia	Legal right exists, but not available in practice
Slovakia	No age limitation
Slovenia	Access after age of 18; before 18 rights in educational settings and other settings assigned by law
Spain	Often only post-compulsory education (varies by region)
Sweden	No age restriction
Switzerland-Fr.	From compulsory schooling (no fixed age) but approved by the Invalidity Insurance
Switzerland-It.	From start of school
Ukraine[HI]	N/A

Across Europe, multiple stakeholders are pushing for action to ensure the legal right to an interpreter. The respondents were asked who in their country or region are active in this manner, the results of this question can be found in figure 5.10. As was the case in 2016 and 2020, the majority indicates this is done by associations of deaf or deafblind persons. In some countries, as in Malta, France and the Netherlands, multiple stakeholders work together.

Spain reports that since educational responsibilities have been transferred to the autonomous communities, each of which manages this service differently. In some cases, staff are hired directly by the public administration; in others, the service is outsourced to organizations representing deaf people. There are also situations where management is subcontracted to private companies with no knowledge or experience in the profession, which negatively impacts the quality of service and the working conditions of the professionals involved.

In Portugal improvement has been seen over the years:

> After the pandemic, interpretation services have become more valued. In schools and in the justice system, it has been guaranteed for years. In public services, it is still necessary to request it.

In Poland, the use of the legal right to a sign language interpreter is most often initiated by the deaf person themselves. Although public institutions are legally obliged to ensure accessibility, the procedure requires the individual to submit a formal request for accessibility (wniosek o zapewnienie dostępności). Without this request, the institution is not required to provide interpreting services. In some cases, advocacy associations such as the Polish Association of the Deaf (PZG) or the Association of Polish Sign Language Interpreters (STPJM) support individuals in submitting such requests or intervene when accessibility is denied. However, in most situations the deaf person must take the first step to activate their legal right to an interpreter.

Fig. 5.10 Who pushes mostly for action to ensure the use of the legal right to an interpreter?

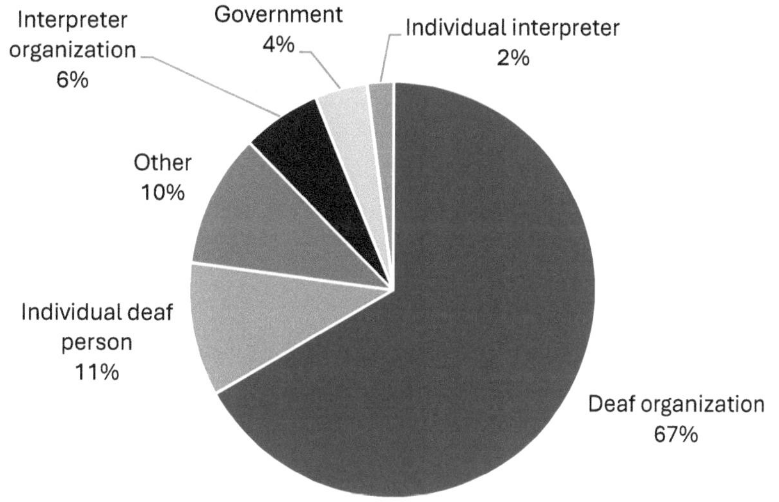

5.9 Requesting interpreting services

There are different ways a deaf person can request the services of a sign language interpreter (Fig. 5.11). Many countries have more than one option.

In some countries there is only one possibility to request an interpreter, for example in Switzerland, where most the requests are handled by Procom. A small group of independent interpreters also manage their own requests through their own website. However, they often receive fewer assignments, as most contracts are centralized through Procom. Iceland has also a centralized service, supplemented by some freelancers.

In Turkey there is an application that provides remote video interpreting services. This service is offered free of charge directly by the government, allowing deaf citizens to receive interpreting support for their daily communication needs. To access the service, users can register through the official website or download the application to their phones. The name of the application is AİLEM Accessible Communication Center. In Slovenia there is a call center established at the association which a deaf person can contact 365 days a year, at any time.

In Montenegro there is no service available to request an interpreter.

Fig. 5.11 Options to request a sign language interpreter (47 countries & regions)

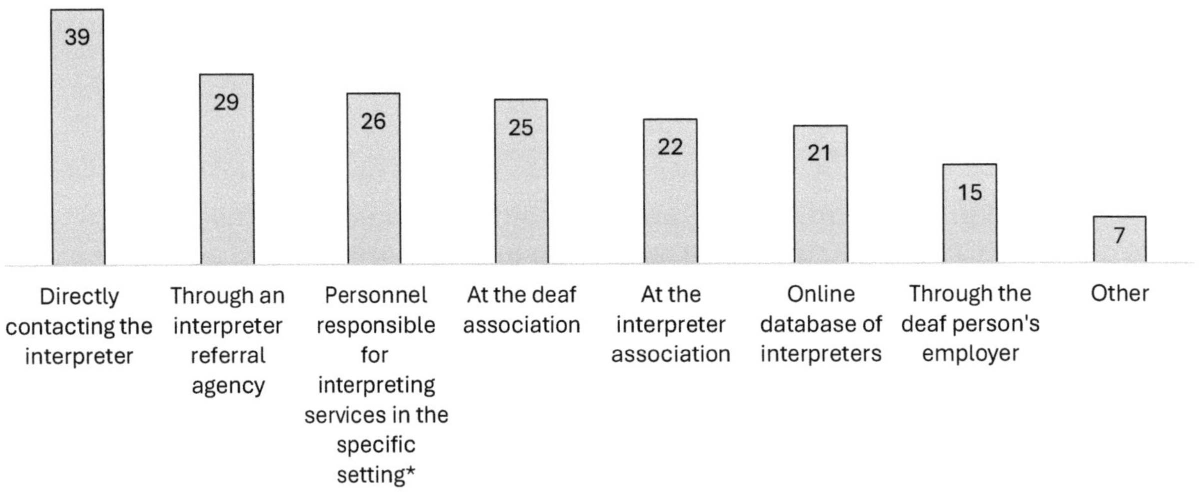

**Hospital, university, etc.*

Most countries or regions allow the deaf person to request a preferred interpreter. Table 5.5 gives an overview in which country or region a deaf or deafblind person can always request their preferred interpreter, and in which country or region the deaf or deafblind person can request a preferred interpreter for a specific setting only, or not at all.

Table 5.5 Possibility for a deaf, hard of hearing or deafblind person to request an interpreter in a specific setting			
Country / region	**Deaf / hard of hearing person**	**Deafblind person**	**Comments:**
Albania	Yes, in certain settings	No	
Austria	Yes, always	Yes, always	
Belgium - Flanders	Yes, always	Yes, always	
Belgium - Wallonia	Yes, always	Yes, always	
Bosnia-Herzegovina	Yes, in certain settings	No	
Bulgaria	Yes, in certain settings	Yes, in certain settings	Due to limited availability of interpreters this is not always possible, but the choice of the deaf person is always considered.
Croatia	Yes, in certain settings	Yes, in certain settings	
Cyprus	Yes, always	Yes, always	
Czech Republic	Yes, in certain settings	Yes, in certain settings	Deafblind – don't know They have the right to choose, but it depends on the capacity and availability.
Denmark	Yes, in certain settings	Yes, in certain settings	
Engl. W&NI	Yes, in certain settings	Yes, in certain settings	
Estonia	Yes, in certain settings	Yes, in certain settings	Yes, the client can choose, but the number of interpreters is so limited, it's not always possible. The client cannot choose the interpreter in remote interpreting.
Finland	Yes, always	Yes, always	They can always request an interpreter from their personal interpreter list, but the interpreter requested might not be available.
France	Yes, in certain settings	Yes, in certain settings	It depends on who makes the reservation. But most of the time, Deaf/Deafblind persons can ask for a preferred interpreter. However, this is not the case in settings like conferences.
Georgia	Yes, in certain settings	No	
Germany	Yes, in certain settings	Yes, in certain settings	
Greece	Yes, always	Yes, always	
Hungary	Yes, always	Yes, always	
Iceland	Yes, always	Yes, always	We try to fulfill - not always possible.
Ireland	Yes, always	Yes, always	
Italy	Yes, in certain settings	Yes, in certain settings	
Kosovo	Yes, always	Yes, always	
Latvia	Yes, in certain settings	Yes, in certain settings	
Lithuania	Yes, always	No	There are not professional deaf blind interpreters in LT.
Luxembourg	Yes, in certain settings	No	It depends on the setting. It can be requested but we are not always available or allowed to work in specific settings.
Malta	Yes, in certain settings	Yes, in certain settings	If available. Not usually requested. We try to follow up on a case so they don't must start from scratch with a new interpreter of and when possible.
Moldova	Yes, in certain settings	No	
Montenegro	No	No	

Table 5.5 Possibility for a deaf, hard of hearing or deafblind person to request an interpreter in a specific setting

Country / region	Deaf / hard of hearing person	Deafblind person	Comments:
Netherlands	Yes, in certain settings	Yes, in certain settings	Not in criminal and immigration proceedings.
Norway	Yes, always	Yes, always	They can always request; the request is not always adhered to.
Poland	No	No	In practice, it is easiest for institutions when a deaf or deafblind person indicates their preferred interpreter, and this is often informally accepted. However, the law does not require institutions to respect that choice. Legally, the responsibility is only to ensure communication accessibility, not to guarantee a specific interpreter. As a result, while personal preference is sometimes considered—especially when the request is submitted early or when the interpreter is known to the institution—it is not a legal right and depends on the institution's willingness and logistical possibilities.
Portugal	Yes, in certain settings	Yes, in certain settings	
Romania	No	No	
Russia	Yes, in certain settings	Yes, in certain settings	
Scotland	Yes, in certain settings	Yes, in certain settings	
Serbia	Yes, always	Yes, always	
Slovakia	Yes, in certain settings	Yes, in certain settings	
Slovenia	Yes, always	Yes, always	When it comes to personal matters, always, when it comes to the client who is the payer, it depends on the client whether to allow this to the deaf person and as a rule, it is allowed. The exception is the court, the police, all urgent cases, where it is not possible to choose who will interpret, but it is more important that they have an interpreter at their disposal, and they also cannot choose one in the media, TV or various streaming channels.
Spain	No	No	The manner in which the service is provided is subject to the criteria and decisions of the entity that manages it.
Sweden	Yes, in certain settings	Yes, in certain settings	
Switzerland – Fr.	Yes, always	Yes, always	Through the interpreting request form, Deaf, hard of hearing, and deafblind clients can indicate their preferences, choosing certain interpreters or declining others.
Switzerland – Ge.	Yes, always	Yes, always	They can always wish, but it is not a guarantee that the interpreter is available.
Switzerland – It.	Yes, always	Yes, always	
Turkey	No	No	The limited number of qualified interpreters and the restricted availability of free interpreting services limit the right of deaf individuals to choose their preferred interpreter.
Ukraine[HI]	Yes, in certain settings	No	

5.10 Party responsible for the payment of interpreting services

Each country in Europe has different regulations and laws on sign language interpreting. The payment of the interpreting services differs accordingly. The respondents were asked who is the paying party responsible for five main settings where interpreting services can be provided: employment (Table 5.6), education (Table 5.7), medical (Table 5.8), legal (Table 5.9), and theatre (Table 5.10).

In most of the countries the government is responsible for paying the interpreting services in employment, educational, medical, and legal settings. The only exception are theatres; in most countries the theatres, the deaf club, or the deaf persons themselves pay for the interpreting services. In educational settings there are also many countries where the educational institutions pay for the interpreting services.

In many countries there are restrictions on the government funding for interpreting services. The restriction is often linked to the situation and, or to a set budget per person. In the Netherlands, for example, deaf people have a right to an interpreter for fifteen percent of their work time, and one hundred percent in education. In addition, they have thirty hours a year for other situations. Deafblind people have a right to a total of 168 hours of interpreting services per year.

Table 5.6 WORK SETTING: Party responsible for the payment of the interpreting services								
Country / region	**Government/ agency**	**Employer**	**Educational program**	**Theatre**	**Deaf Club**	**Deaf person**	**Other**	**N/A**
Albania		X						
Austria	X	X						
Belgium-Flanders	X	X						
Belgium-Wallonia	X	X						
Bosnia-Herzegovina			X					
Bulgaria	X					X		
Croatia	X	X						
Cyprus	X	X						
Czech Republic	X	X						
Denmark	X							
Engl. W&NI	X							
Estonia	X	X						
Finland	X							
France		X						
Georgia								X
Germany	X	X						
Greece	X	X	X					
Hungary	X	X				X		
Iceland	X							
Ireland	X							
Italy - ANIMU		X						
Italy - ANIOS	X	X						
Kosovo	X	X						
Latvia	X							
Lithuania	X							
Luxembourg	X						X	
Malta	X							
Moldova	X	X						
Montenegro								X
Netherlands								X
Norway	X							
Poland	X	X	X	X	X	X		
Portugal	X	X			X	X		
Romania							X	
Russia		X						
Scotland		X					X	
Serbia					X	X		
Slovakia	X							
Slovenia	X	X						
Spain		X						
Sweden	X	X						
Switzerland - Fr.	X							
Switzerland - Ge.	X	X						
Switzerland - It.	X							
Turkey	X	X			X			
Ukraine[HI]		X				X		

Table 5.7 EDUCATIONAL SETTING: Party responsible for the payment of the interpreting services								
Country / region	**Government/ agency**	**Employer**	**Educational program**	**Theatr e**	**Deaf Club**	**Deaf person**	**Other**	**N/A**
Albania	x							
Austria	x		x					
Belgium-Flanders	x							
Belgium-Wallonia	x							
Bosnia-Herzegovina			x		x	x		
Bulgaria	x		x			x		
Croatia	x							
Cyprus	x							
Czech Republic	x		x					
Denmark	x							
Engl. W&NI	x		x					
Estonia	x		x			x		
Finland	x							
France			x					
Georgia								x
Germany	x							
Greece			x					
Hungary	x					x		
Iceland	x							
Ireland	x							
Italy - ANIMU	x		x					
Italy - ANIOS	x				x			
Kosovo	x							
Latvia	x							
Lithuania	x							
Luxembourg	x							
Malta	x							
Moldova						x		
Montenegro								x
Netherlands								x
Norway	x							
Poland	x		x					
Portugal	x		x					
Romania							x	
Russia		x						
Scotland	x							
Serbia								x
Slovakia	x							
Slovenia	x	x				x		
Spain	x	x						
Sweden	x		x					
Switzerland-Fr.	x							
Switzerland-Ge.	x							
Switzerland-It.	x							
Turkey	x	x	x		x			
Ukraine[HI]	x		x		x	x		

Table 5.8 MEDICAL SETTING: Party responsible for the payment of the interpreting services

Country / region	Government/ agency	Employer	Educational program	Theatre	Deaf Club	deaf person	Other	N/A
Albania								x
Austria	x							
Belgium-Flanders	x						x	
Belgium-Wallonia	x							
Bosnia-Herzegovina						x		
Bulgaria	x					x		
Croatia	x							
Cyprus	x							
Czech Republic	x							
Denmark	x							
Engl. W&NI	x							
Estonia	x							
Finland	x							
France	x					x		
Georgia								x
Germany	x							
Greece					x			
Hungary	x					x		
Iceland	x							
Ireland	x							
Italy - ANIMU	x					x	x	
Italy - ANIOS	x				x			
Kosovo	x							
Latvia	x							
Lithuania	x							
Luxembourg						x	x	
Malta	x							
Moldova	x	x						
Montenegro								x
Netherlands								x
Norway	x							
Poland	x					x		
Portugal	x					x		
Romania						x		
Russia	x							
Scotland	x							
Serbia					x			
Slovakia	x							
Slovenia	x							
Spain		x						
Sweden	x							
Switzerland-Fr.						x		
Switzerland-Ge.	x							
Switzerland-It.	x							
Turkey	x							
Ukraine[HI]	x				x	x		

Table 5.9 LEGAL SETTING: Party responsible for the payment of the interpreting services								
Country / region	**Government/ agency**	**Employer**	**Educational program**	**Theatre**	**Deaf Club**	**deaf person**	**Other**	**N/A**
Albania	x							
Austria	x							
Belgium-Flanders	x						x	
Belgium-Wallonia	x							
Bosnia-Herzegovina	x					x		
Bulgaria	x					x		
Croatia	x							
Cyprus	x							
Czech Republic	x							
Denmark	x							
Engl. W&NI	x					x		
Estonia	x							
Finland	x							
France	x					x		
Georgia								x
Germany	x							
Greece		x			x			
Hungary	x					x		
Iceland	x							
Ireland	x							
Italy - ANIMU	x					x		
Italy - ANIOS	x				x	x		
Kosovo	x							
Latvia	x							
Lithuania	x							
Luxembourg	x							
Malta	x							
Moldova	x	x				x		
Montenegro								x
Netherlands								x
Norway	x							
Poland	x							
Portugal	x							
Romania	x					x		
Russia	x					x		
Scotland	x							
Serbia	x					x		
Slovakia	x							
Slovenia	x							
Spain		x						
Sweden	x							
Switzerland-Fr.	x							
Switzerland-Ge.	x							
Switzerland-It.	x							
Turkey	x							
Ukraine[HI]	x				x	x		

Table 5.10 THEATRE SETTING: Party responsible for the payment of the interpreting services								
Country / region	**Government/ agency**	**Employer**	**Educational program**	**Theatre**	**Deaf Club**	**deaf person**	**Other**	**N/A**
Albania								x
Austria				x	x	x		
Belgium-Flanders	x			x				
Belgium-Wallonia	x					x		
Bosnia-Herzegovina			x	x	x		x	
Bulgaria	x							
Croatia	x							
Cyprus	x							
Czech Republic				x	x		x	
Denmark	x							
Engl. W&NI				x				
Estonia	x			x				
Finland	x			x				
France				x				
Georgia								x
Germany				x			x	
Greece		x		x	x			
Hungary				x				
Iceland				x				
Ireland				x				
Italy - ANIMU				x				
Italy - ANIOS				x	x	x		
Kosovo								x
Latvia				x				
Lithuania	x							
Luxembourg	x						x	
Malta				x		x	x	
Moldova	x	x			x			
Montenegro								x
Netherlands								x
Norway	x							
Poland	x			x				
Portugal	x			x				
Romania				x				
Russia	x			x			x	
Scotland				x				
Serbia				x	x			
Slovakia				x			x	
Slovenia						x		
Spain				x				
Sweden	x							
Switzerland-Fr.				x	x		x	
Switzerland-Ge.				x				
Switzerland-It.						x		
Turkey					x			
Ukraine[HI]	x			x	x			

5.11 Recognition of sign language

Except for Switzerland, all countries and regions have officially recognized their national sign language(s) as an official language. An overview of the countries, sign languages, and the year of recognition can be found in Table 5.11. The first to recognize sign language as an official language was Sweden in 1981. The most recent one was Norway in 2022. Since the last survey in 2020, three more countries have recognized sign language: Bulgaria, Italy and Norway. All the 27 EU member states, have now legally recognized their national sign language as a language .

In Switzerland the cantons of Zürich, Geneva, Ticino and Neuchâtel have recognized the signed languages in their constitutions or in their legislation. The Confederation has not concretely recognized sign languages, and the cantons have not taken concrete measures either to act on this recognition.[97]

Table 5.11 Recognition of sign language per country or region

Country / region	Year of recognition	Name of the sign language
Sweden	1981	Svenskt teckenspråk (SSL)
Finland	1995	Suomalainen viittomakieli & Finlandssvenskt
Lithuania	1995	Lietuviu gestu kalba
Slovakia	1995	Slovenský posunkový jazyk
Portugal	1997	Intérpretes de Língua Gestual Portuguesa (LGP)
Czech Republic	1998	Česky znakovy jazyk (CZJ)
Latvia	2000	Latviešu zīmju valoda
Greece	2000	Ελληνική Νοηματική Γλώσσα – Greek Sign Language (ΕΝΓ)
Germany	2002	Deutsche Gebärdensprache (DGS)
Romania	2002	Limbaj Mimico-Gestual Romanesc (LMGR)
Slovenia	2002	Slovenski znakovni jezik
Belgium-Wallonia	2003	Langue des Signes de Belgique Francophone (LSFB)
England, Wales & NI	2003,	British Sign Language (BSL)
Austria	2005	Österreichische Gebärdensprache
France	2005	Langue des Signes Française (LSF)
Turkey	2005	Türk İşaret Dili (TID)
Belgium-Flanders	2006	Vlaamse Gebarentaal (VGT)
Cyprus	2006	Κυπριακή Νοηματική Γλώσσα - Cypriot Sign Language (ΚΝΓ)
Estonia	2007	Eesti viipekeel (EVK)
Spain	2007	Lengua de Signos Espanola & Lengua de Signos Catalna (LSE
Hungary	2009	Magyar jelnyelv
Kosovo	2010	Gjuha e Shenjave Kosovare
Iceland	2011	Islenskt táknmál (ÍTM)
Ukraine	2004	Українська жестова мова - Ukrainian Sign Language
Poland	2011	Polski Jezyk Migowy (PJM)

[97] https://www.sgb-fss.ch/fr/domaines/travail-politique/reconnaissance-de-la-langue-des-signes/

Table 5.11 Recognition of sign language per country or region

Country / region	Year of recognition	Name of the sign language
Armenia	2011	Armenian Sign Language
Russia	2012	Russkij zhestovyj yazyk (русский жестовый язык)
Denmark	2014	Dansk Tegnsprog
Albania	2014	Gjuha e Shenjave Shqipe
Serbia	2015	Srpski znakovni jezik
Scotland	2015	British Sign Language (BSL)
Croatia	2015	Hrvatski znakovni jezik
Malta	2016	Lingwa Sinjali Maltija (LSM)
Ireland	2017	Irish Sign Language (ISL)
Bosnia & Herzegovina	2018	Znakovni jezik gluvih
Luxembourg	2018	Deutsche Gebärdensprache (DGS)
Georgia	2020	ქართული ჟესტური ენა - Georgian Sign Language
The Netherlands	2020	Nederlandse Gebarentaal (NGT)
Bulgaria	2021	Български жестов език (БЖЕ) - Bulgarian Sign Language
Italy	2021	Lingua dei segni italiana (LIS)
Norway	2022	Norsk tegnspråk (NTS)
Switzerland	Not recognized at federal level. Recognized by the cantons of Zürich,	Lingua italiana dei segni (LIS) Langue des signes française (LSF) Deutschschweizer Gebärdensprache (DSGS)

5.12 Forms of employment

Sign language interpreters across Europe work in different forms of employment (Fig. 5.12). Most sign language interpreters work as freelancer in combination with working for an agency or employer and as freelancer. This has changed since 2020, when they mostly worked for an agency and some as freelancers. It seems there is a trend towards freelancing. Table 5.12 shows the forms of employment per country.

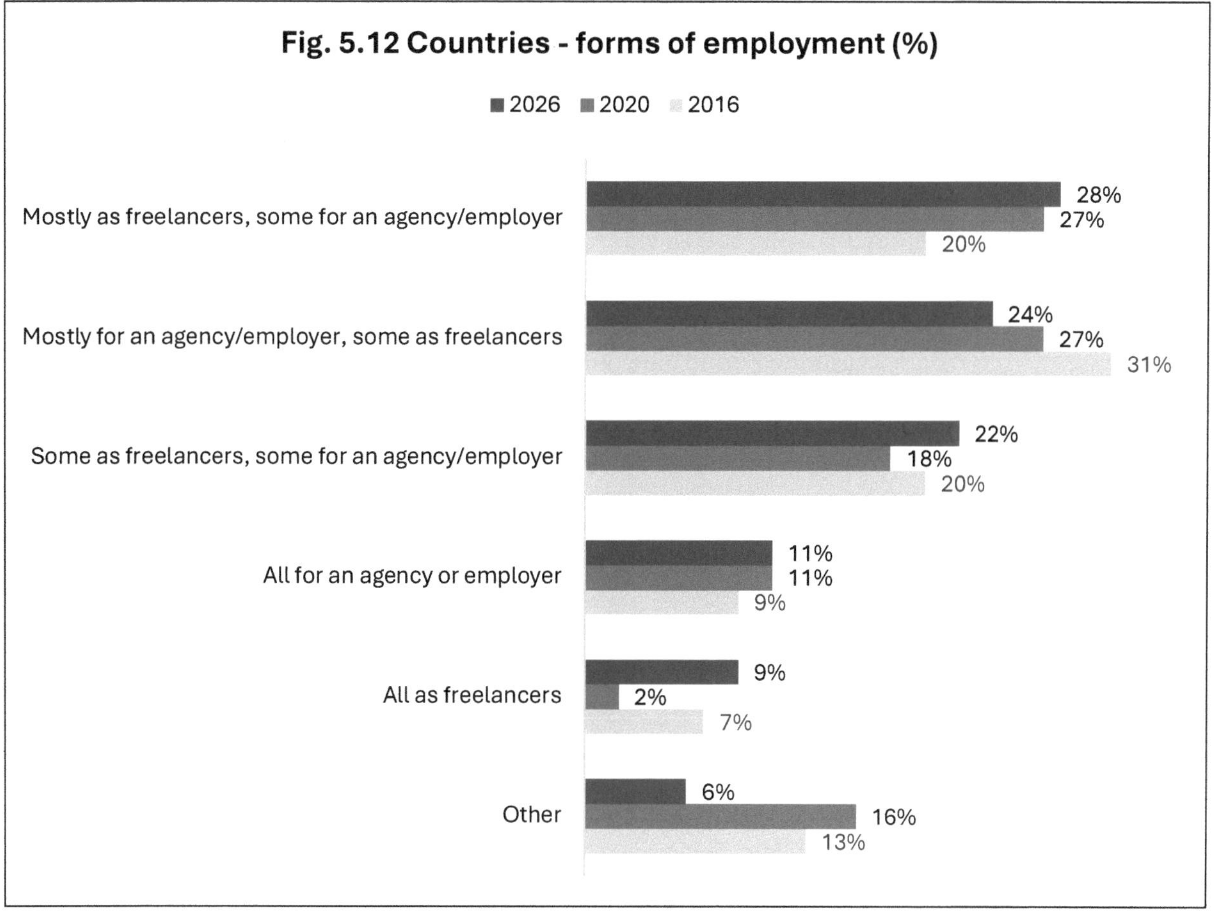

Table 5.12 Employment forms per country and region

All as freelancers	Mostly as freelancers, some for an agency	50% freelancers, 50% for an agency/ employer	Mostly for an agency/employer, some as freelancers	All for an agency/employer
Albania	Austria	Belgium-Wallonia	Czech Republic	Croatia
Bosnia-Herzegovina	Belgium-Flanders	Cyprus	Iceland	Estonia
Kosovo	Bulgaria	Finland	Lithuania	Latvia
Slovenia	Denmark	France	Malta	Luxembourg
	Engl. W&NI	Hungary	Moldova	Switzerland-It.
	Germany	Noray	Serbia	
	Greece	Portugal	Spain	
	Ireland	Russia	Sweden	
	Italy	Slovakia	Switzerland-Fr.	
	Netherlands	Turkey	Switzerland-Ge.	
	Poland		Ukraine[HI]	
	Romania			
	Scotland			

5.13 Onsite and remote interpreting

Following the Covid19 pandemic, new questions to this survey have been added on remote and onsite interpreting. As Figure 5.13 shows that there is mostly a combination of onsite and remote work (64%). Additionally, Figure 5.14 shows that the requests for remote interpreting have significantly increased following the Covid19 pandemic. Not every country reports a change. For example, in Iceland, Kosovo, Montenegro, Slovakia, and Slovenia there was no change in the remote interpreting requests. The positive change is that most countries show an improvement in the remote working conditions (Fig. 5.15).

Fig. 5.13 How interpreters mostly work in the countries (n=47)

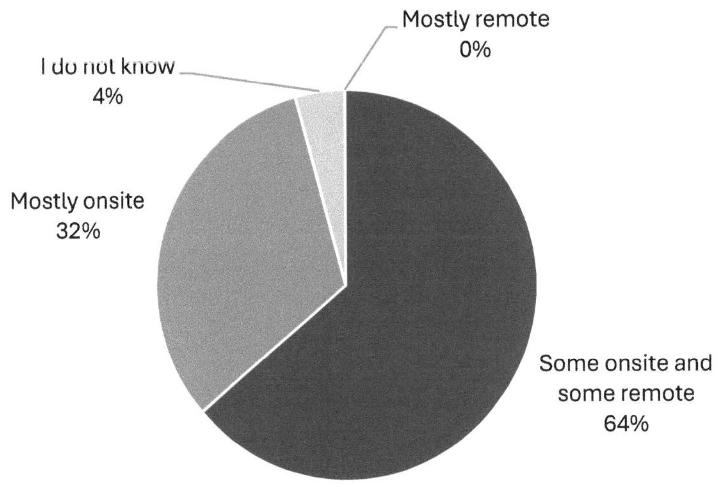

Fig. 5.14 Since the Covid19 pandemic changes in the requests for remote interpreting (n=47)

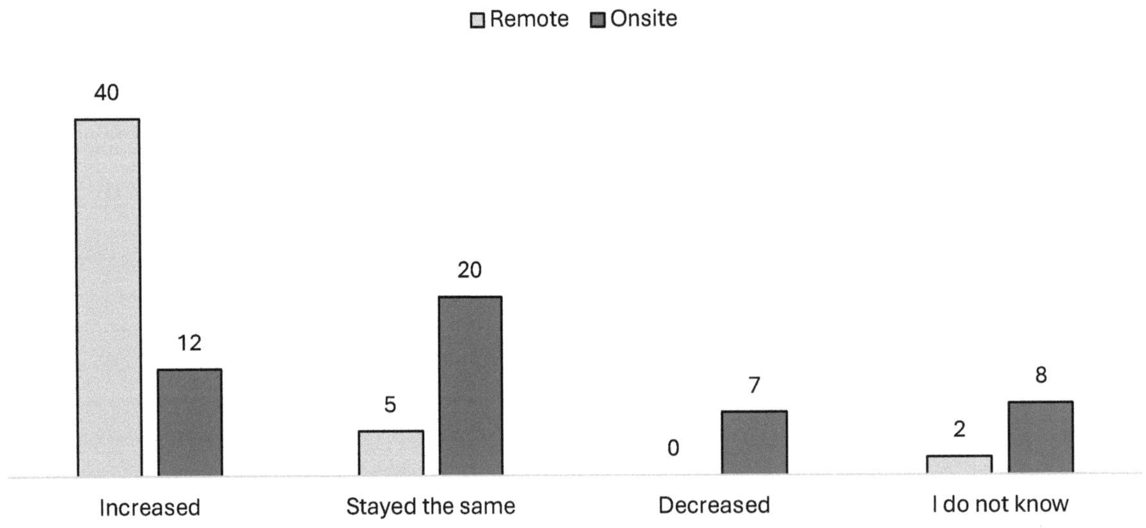

Fig. 5.15 Change in working conditions for remote interpreting since the Covid19 pandemic (n=46)

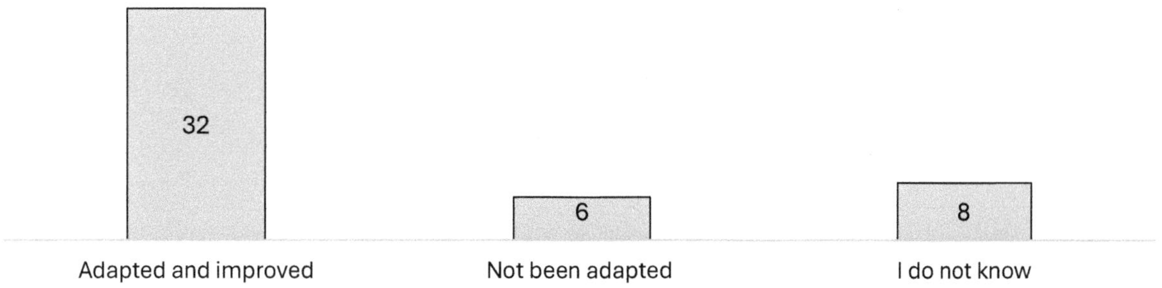

5.14 Freelance interpreting fees

Fees for freelance work as a sign language interpreter vary per country and region. Not all respondents are able to provide an answer as to how interpreters are employed or paid in their country. The map in Figure 5.16 provides a geographical overview of the minimum freelance interpreting fee per hour.

The survey question contained four sections: interpreting fee per hour, compensation for traveling time, traveling expenses, and other expenses. The lowest fee per hour is € 3.00 in Georgia and in Russia, and the highest fee is € 130 per hour in the French and German regions of Switzerland. When respondents provided a range in the fee per hour, the lowest fee was used in both Figures (5.16, 5.17) and the numbers were rounded up. For example, in Belgium-Flanders the minimum fee per hour is € 56.75 per hour, in Figure 5.17 it shows € 57.00 per freelance hour. Table 5.13 provides an overview of the exact numbers, as well as all the payment entitlements.

A freelancer can receive additional benefits, for example social security and pension entitlements, in some of the countries: Austria, Denmark, Estonia, Finland, Slovenia, Sweden, and Switzerland-Italian region. In most countries the interpreters need to manage their own social security and pension provisions.

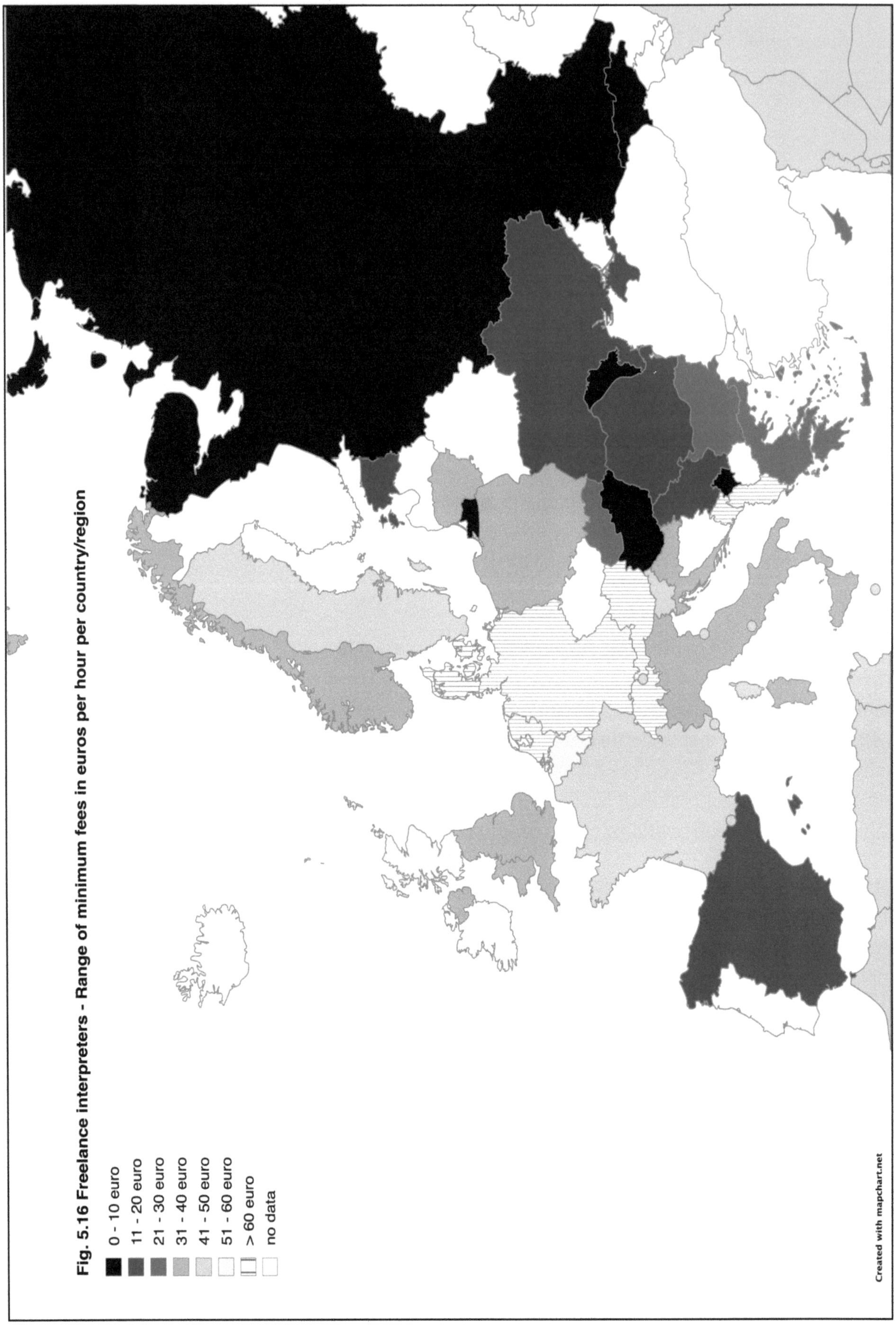

Fig. 5.16 Freelance interpreters - Range of minimum fees in euros per hour per country/region

Fig. 5.17 Minimum interpreting fee per freelance hour in 2025*

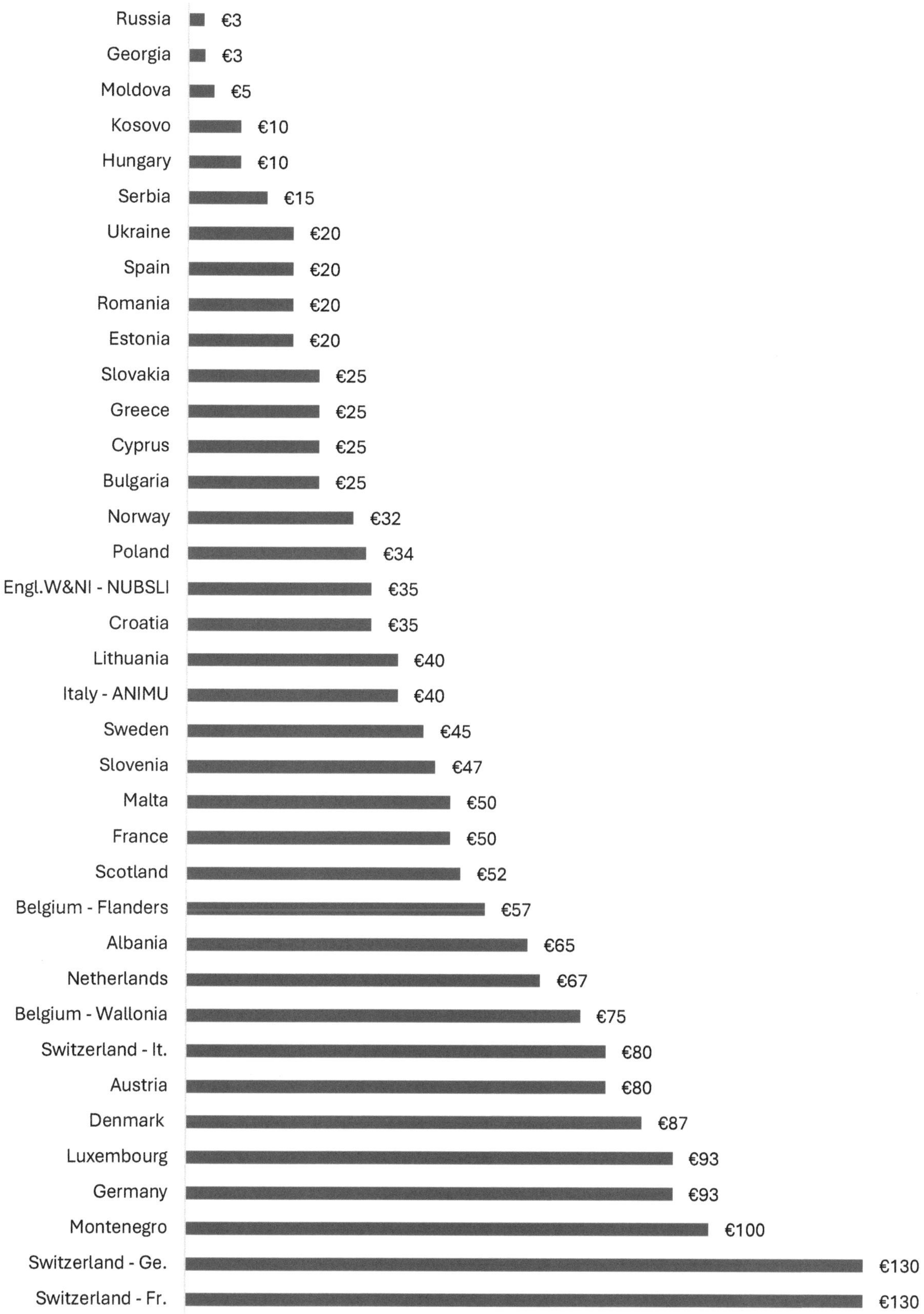

**The graph does not include all countries and regions. It only includes those that have an hourly fee and were able to provide the data. The fees are rounded off and are <u>minimum</u> fees. See Table 5.13 for detailed answers.*

Table 5.13 Overview per country of payment for freelance interpreting

Country / region	Fee per hour	Travel time	Travel expenses	Increase fee with interpreting experience	Minimum fee or minimum time per assignment?
Albania	€ 65	0		no	yes
Austria	€ 80	€ 35	Depending on mode of transportation.	yes	no
Belgium - Flanders	€ 56.76	0	Travel fee per km changes every 3 months (currently € 0.4309 per km).	no	yes
Belgium - Wallonia	€ 75	Depends, an hour or nothing.	€ 0.50 per km or public transport.	yes	no
Bosnia-Herzegovina*				no	no
Bulgaria	€ 25	0	0	no	yes
Croatia	€ 35	Depends, usually not covered.	Depends, usually fully covered.	no	no
Cyprus	€ 25	€ 0.1709 per kilometer		no	yes
Czech Republic	DK			yes	no
Denmark	€ 87 - € 100	0	0	no	yes
Engl. W&NI - ASLI	0	0	Full	no	yes
Engl. W&NI - NUBSLI	€ 35	€ 0.5	N/A	yes	yes
Estonia	€ 20 - € 50 depending on the proficiency and the assignment	Half of the interpreting fee.	Whatever the expense is.	yes	yes
Finland	DK			yes	yes
France	€ 50 - € 70 for liaison, € 80 - € 100 for meetings and conferences		It is usually calculated on the real price.	not sure	yes
Georgia	under the state program € 3,18 (10 lari)				no
Germany	€ 93	€ 75	€ 0.35 per km	no	yes
Greece	€ 25 - € 60	0	100	yes	yes
Hungary	€ 10	0	0	no	
Iceland	DK	DK	DK	yes	no
Ireland	interpreters charge a fee, not an hourly rate. Average minimum fee is € 145 - € 150 (https://cisli.ie/resources/fees-surveys/).	Not often charged/paid.	Normally per mileage.	no	yes
Italy - ANIMU	€ 40 - € 80 depending on the region			no	no
Italy - ANIOS				no	no
Kosovo	€ 10	€ 10	0	no	yes
Latvia				yes	yes
Lithuania	€ 40	0	0	no	no
Luxembourg	€ 93	€ 0.42 per km (travel and waiting times).		yes	yes
Malta	€ 50	NA	NA	no	yes
Moldova	€ 4.89	0	0	no	yes
Montenegro	€ 100	0	DK	no	yes
Netherlands	€ 67.40 for educational settings, € 77.24 for work and community settings.	€ 0.87 per km (incl. traveling time and expenses).	€ 0.87 per km (incl. traveling time and expenses).	no	yes

Table 5.13 Overview per country of payment for freelance interpreting

Country / region	Fee per hour	Travel time	Travel expenses	Increase fee with interpreting experience	Minimum fee or minimum time per assignment?
Norway	€ 31.57	€ 18.97	Traveling expenses are covered.	no	yes
Poland	Approximately € 34 (around 150 PLN). Interpreters who interpret part-time or as a secondary occupation (about 70% of the profession) often agree to lower rates, averaging € 27–€ 28 (around 120 PLN) per hour.	Interpreters are not paid for travel time.	Within the same city, travel costs are not reimbursed. For travel to another city—particularly for distances exceeding 200 km—travel expenses are negotiated individually, and in some cases, interpreters receive full or partial reimbursement under the condition that otherwise they would decline the assignment.	yes	no
Portugal	DK	€ 30 – € 45 per hour	Payment of travel expenses is requested from the employer.	yes	no
Romania	€ 20 - € 100	Depends.	Depends.	yes	no
Russia	Depends which region - overall the governmental rate per hour is € 3. Though Moscow rates would start from € 10 up to € 50	0	0	no	no
Scotland	€ 52	Usually negotiated hourly rate.	€ 0.50 - € 0.60 per mile		
Serbia	Freelance interpreters in Serbia earn on average between 1,800 and 12,000 RSD per hour (approximately € 15 – € 100 per hour, depending on the assignment)	0	0	no	no
Slovakia	€ 25			yes	no
Slovenia	€ 47.19	/	€ 0.18 km /public transfer all	no	yes
Spain	€ 20 - € 70	NA	€ 0.20 per km, toll and parking if applicable.	yes	no
Sweden	~ € 45	The tenders are too different to get an average.	~22	no	no
Switz. - Fr.	€ 130	€ 40	public transport or € 0.80 per km.	yes	yes
Switz. - Ge.	€ 130	€ 50		yes	yes
Switz. - It.	€ 80	€ 40	€ 20	yes	yes
Turkey	There is no standard on this matter.		Sometimes the interpreter covers the cost, sometimes the deaf person does, and in some cases, the employer provides the funding.	yes	no
Ukraine[HI]	€ 20 - € 75	0	In most cases, the customer pays for travel expenses or provides tickets for travel.	yes	yes

** Based on an individual response*

DK=Don't know

5.15 Payment for specific settings or competencies

Interpreters can also get extra payment for specialties and specific settings (Table 5.14). Figure 5.18 shows that this is most common in conference settings, TV and media, as well as when the interpreter knows additional languages. As interpreters are increasingly working remotely, there is also a possibility to receive extra payment for remote interpretation and web streaming, or the recording of the interpretation and copyright fees (Fig. 5.19).

Table 5.14 Extra pay for specialties and specific settings

Country / region	Legal (court, police, etc.)	Healthcare - general	Mental health	Conference	Academic	Higher education	Theatre	TV/Media	Deafblind	Additional languages	Other	Remote interpreting	Web streaming	Recording interpretation	Copyright
Albania	no	no	no	yes	no	no	no	no	no	no	no	no	no	no	no
Austria	NA	NA	NA	yes	yes	yes	yes	yes	NA	yes	NA	NA	yes	yes	yes
Belgium - Flanders	yes	yes	yes	yes	no	no	yes	yes	yes	no	yes	yes	yes	yes	yes
	Only applicable if the interpreter negotiates the fee themselves. In many cases it is paid for by the government which is always a fixed fee (no exceptions/extras).														
Belgium – Wallonia	no	no	no	yes	no	no	no	no	no	yes	NA	yes	yes	yes	yes
Bulgaria	yes	no	NA	yes	NA	NA	no	no	no	NA	no	no	no	yes	NA
Croatia	yes	no	no	yes	no	no	yes	yes	yes	yes	NA	no	yes	yes	yes
Cyprus	no	no	no	no	no	no	no	no	no	no	no	no	no	no	no
Czech Republic	NA	NA	NA	yes	NA	NA	NA	yes	NA	yes	NA	NA	yes	NA	NA
Denmark	yes	no	no	no	no	no	yes	yes	yes	yes	NA	no	no	no	no
Denmark[DI]	yes	no	no	yes	no	no	yes	yes	yes	yes	NA	no	no	no	no
Engl. W&NI-ASLI	NA	no	no	no	no	no	no	no	no	no	NA	no	no	yes	NA
	Very varied across the profession.														
Engl. W&NI-NUBSLI	yes	yes	yes	yes	yes	yes	yes	yes	yes	yes	yes	yes	yes	yes	yes
	Depends on the individual interpreter.														
Estonia	yes	yes	yes	yes	yes	yes	yes	yes	yes	yes	yes	yes	no	no	no
	The main employers have different fees for all above mentioned assignments, so, when the event is streamed or recorded, already a higher fee is usually applied.														
Finland	yes	no	yes	yes	yes	yes	yes	yes	yes	yes	yes	no	no	yes	NA
France	no	no	no	yes	no	no	no	yes	no	yes	NA	yes	yes	yes	yes
	It depends on who is paying. Legal interpreting: the government decides on how much interpreters (no matter the language) are paid and its around € 40/hour.														
Georgia	yes	yes	yes	yes	yes	yes	yes	yes	yes	yes	yes	yes	yes	yes	yes
	In all fields it is possible to receive an extra payment														
Germany	no	no	no	yes	yes	no	no	no	no	yes	yes	no	yes	yes	yes
Greece	yes	no	no	yes	no	no	yes	yes	NA	yes	no	no	no	yes	yes
Hungary	no	no	no	yes	no	no	NA	yes	no	no	no	no	no	NA	NA
Iceland	no	no	no	no	no	no	no	no	no	yes	no	no	no	no	no

Table 5.14 Extra pay for specialties and specific settings															
Country / region	**Legal (court, police, etc.)**	**Healthcare - general**	**Mental health**	**Conference**	**Academic**	**Higher education**	**Theatre**	**TV/Media**	**Deafblind**	**Additional languages**	**Other**	**Remote interpreting**	**Web streaming**	**Recording interpretation**	**Copyright**
Ireland	yes	yes	yes	yes	no	no	yes	yes	no	NA	NA	no	no	no	no
Italy-ANIMU	no	no	no	yes	yes	no	yes	yes	no	yes	NA	no	yes	yes	yes
Italy-ANIOS	no	no	no	yes	no	no	no	NA	NA	yes	NA	no	yes	yes	yes
Kosovo	no	no	no	yes	no	no	NA	yes	NA	NA	NA	yes	NA	yes	NA
Latvia	yes	no	no	yes	yes	yes	yes	yes	no	yes	yes	yes	yes	yes	yes
Lithuania	NA	NA	NA	NA	NA	NA	NA	NA	NA	NA	NA	NA	NA	NA	NA
	Only for overtime work and increased workload														
Luxembourg	yes	no	no	no	no	no	no	yes	no	yes	no	no	yes	yes	yes
Malta	no	no	no	yes	no	no	yes	yes	NA	NA	NA	no	NA	no	no
Moldova	no	no	no	no	no	no	no	no	no	no	NA	no	no	no	no
Montenegro	yes	NA	NA	NA	NA	NA	NA	NA	NA	NA	NA	NA	NA	NA	NA
	There are only five court interpreters.														
Netherlands	no	no	no	no	no	yes	no	no	no	no	yes	yes	no	no	no
	Remote Interpreting: 130%. Increased fee for after hours, weekends and holidays.														
Norway	yes	no	no	no	no	no	no	no	yes	no	NA	yes	no	no	no
	Remote interpreting: you can be paid a half-hour extra to set up.														
Poland	no	no	no	yes	no	no	yes	no	no	yes	no	no	no	no	no
Portugal	no	no	no	yes	yes	no	no	no	no	NA	NA	no	yes	yes	yes
Romania	no	no	no	no	no	no	yes	no	NA	NA	yes	NA	NA	NA	NA
Russia	no	no	no	yes	no	no	no	yes	no	no	no	no	no	no	no
Scotland	yes	NA	yes	yes	NA	NA	yes	yes	NA	NA	NA	no	no	no	no
Serbia	no	no	no	yes	NA	NA	yes	yes	no	yes	NA	no	no	no	no
Slovakia	no	no	no	no	no	no	no	no	no	no	NA	no	no	no	no
	Only regional governments can provide extra pay when an interpreter is employed full time.														
Slovenia	yes	no	no	no	no	no	no	no	no	yes	yes	no	no	no	yes
Spain	no	no	no	no	no	no	no	NA	no	NA	no	no	no	no	no
	There is usually a standard rate for sign language interpreting services that does not consider the specialization or complexity of the field in which the professional work takes place. For this reason, no additional fees or compensation are paid, even when the interpreting requires a high level of preparation or specific technical knowledge. However, different rates are established in other contexts, such as interpreting between spoken languages or internationally, or in artistic services, such as theatre, where, if the interpreter is part of the cast or participates in multiple performances, the remuneration is adjusted differently.														
Sweden	no	no	no	no	no	no	no	yes	no	yes	NA	no	yes	no	NA
Switz.-Fr.	no	no	no	yes	no	no	yes	yes	no	no	NA	yes	NA	yes	NA
	Web streaming must be mentioned when the request is made. Not sure if Procom or freelance interpreters charge an additional fee for recordings.														
Switz.-Ge.	no	no	no	yes	no	no	yes	no	no	yes	NA	no	no	yes	NA
Switz.- It.	no	no	no	yes	no	no	yes	yes	NA	yes	yes	no	no	yes	yes
Turkey	no	no	no	yes	no	no	yes	yes	NA	yes	NA	no	no	no	no
Ukraine[HI]	yes	no	NA	yes	NA	no	no	yes	NA	yes	NA	no	no	yes	yes

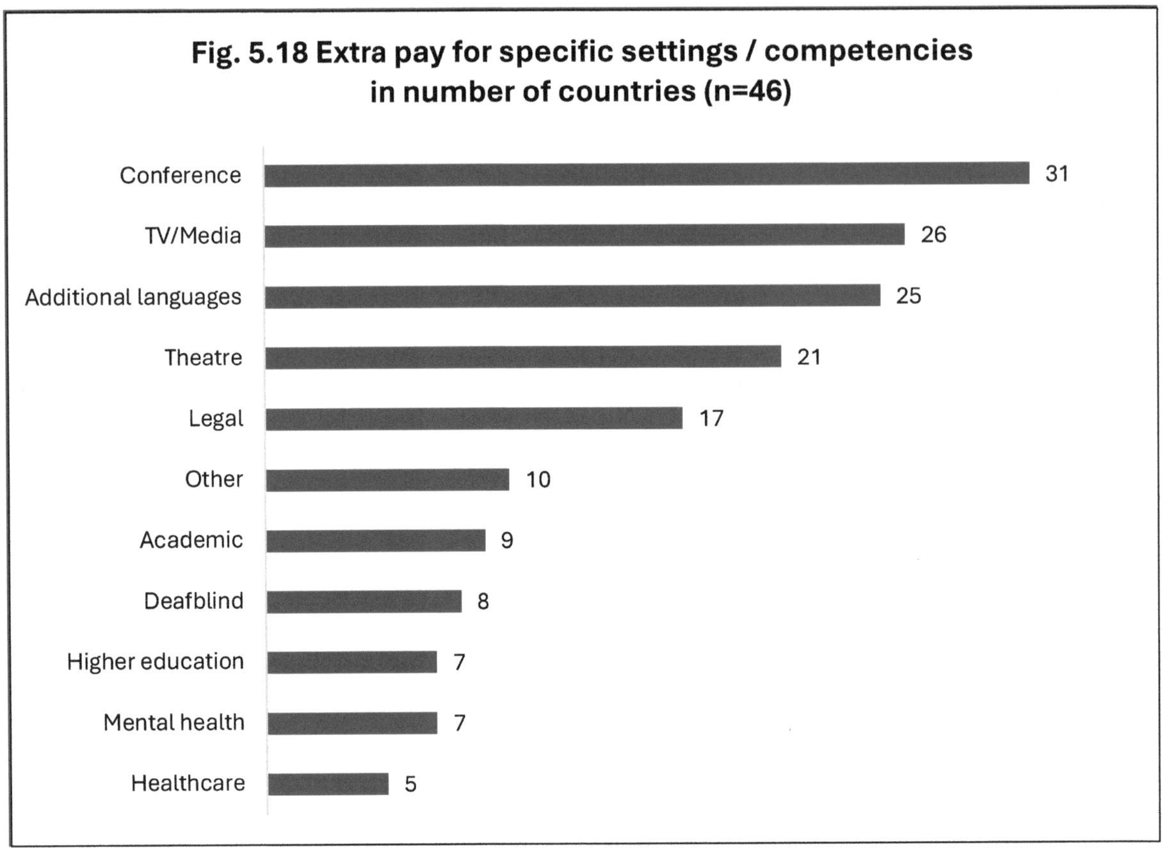

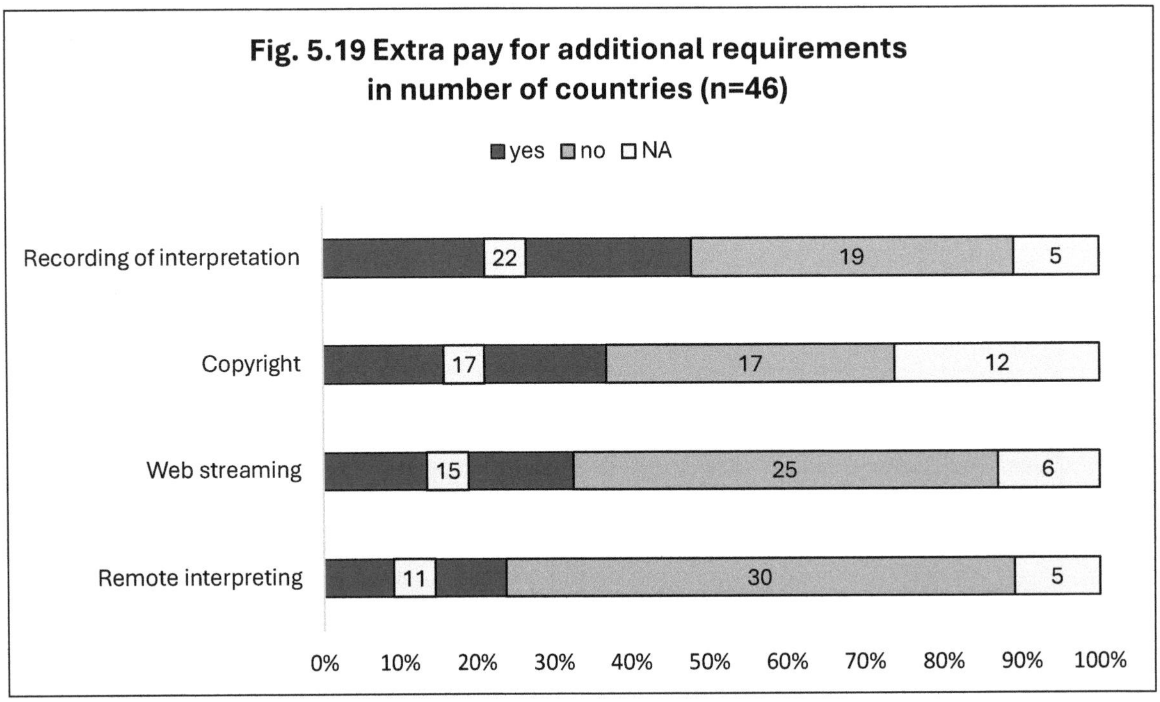

5.16 Interpreting for deafblind persons

When interpreting for deafblind persons, an interpreter sometimes picks up the client at home or does also some guiding. Only in Denmark, the Netherlands, and Russia does the interpreter receive additional remuneration for picking up the deafblind person at home. In fourteen countries interpreters also provide guiding. In very few countries interpreters work in teams of two. The results of this question are presented in Figure 5.20 and detailed per country in Table 5.15.

Fig. 5.20 Interpreter interpreting for a deafblind person (N=47)

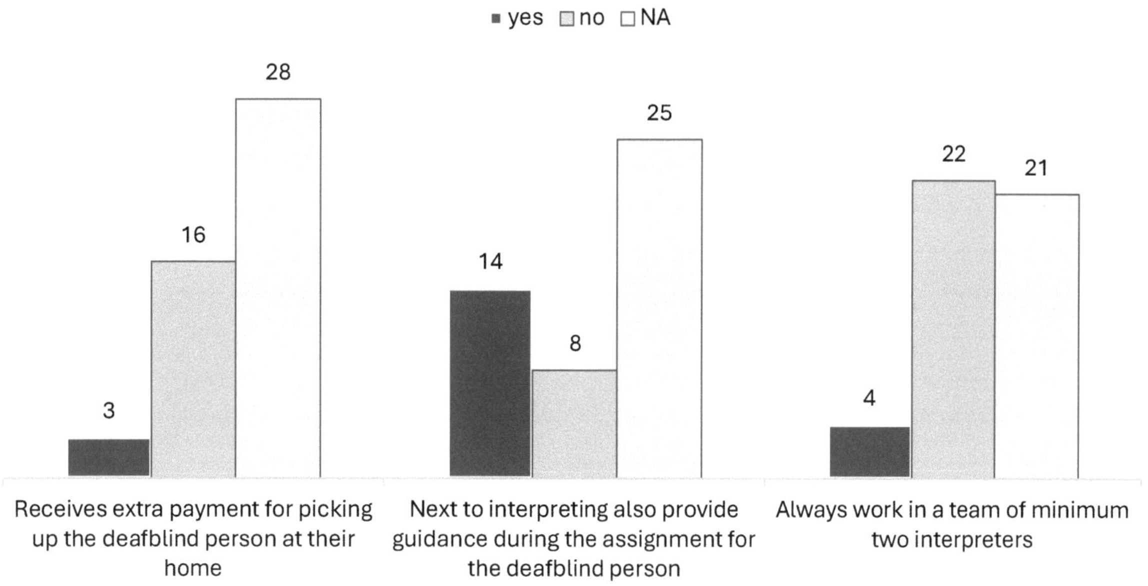

Table 5.15 Interpreting for deafblind persons – the interpreter:			
Country / region	**Receives extra payment for picking up the deafblind person at their home**	**Next to interpreting also provide guidance during the assignment for the deafblind person**	**Always work in teams of minimum two interpreters**
Albania	NA	NA	no
Austria	NA	NA	For assignments of 1 hour or more
Belgium - Flanders	no	yes	no
Belgium - Wallonia	NA	NA	NA
Bosnia-Herzegovina	NA	NA	NA
Bulgaria	no	yes	yes
Croatia	no	yes	no
Cyprus	no	yes	no
Czech Republic	NA	NA	NA
Denmark	yes	no	yes
Engl. W&NI - ASLI	no	NA	no
Engl. W&NI - NUBSLI	NA	NA	NA
Estonia	NA	yes	yes
Finland	NA	NA	NA
France	NA	yes	no
Georgia	NA	NA	NA
Germany	NA	NA	NA

Table 5.15 Interpreting for deafblind persons – the interpreter:			
Country / region	**Receives extra payment for picking up the deafblind person at their home**	**Next to interpreting also provide guidance during the assignment for the deafblind person**	**Always work in teams of minimum two interpreters**
Greece	NA	NA	NA
Hungary	no	no	no
Iceland	no	yes	yes
Ireland	NA	NA	no
Italy - ANIMU	NA	NA	NA
Italy - ANIOS	NA	NA	NA
Kosovo	NA	NA	NA
Latvia	no	no	no
Lithuania	NA	NA	NA
Luxembourg	NA	NA	NA
Malta	NA	yes	no
Moldova	no	no	NA
Montenegro	NA	NA	NA
Netherlands	yes	yes	no
Norway	no	yes	no
Poland	no	yes	no
Portugal	NA	NA	NA
Romania	NA	NA	no
Russia	yes	no	no
Scotland	NA	NA	NA
Serbia	no	no	no
Slovakia	no	yes	no
Slovenia	no	no	no
Spain	no	yes	no
Sweden	no	yes	no
Switzerland-Fr.	NA	NA	no
Switzerland-Ge.	NA	no	no
Switzerland-It.	NA	NA	NA
Turkey	NA	NA	NA
Turkey	NA	NA	NA
Ukraine[HI]	NA	NA	NA

Table 5.16 provides an overview of who determines the fees for freelance interpreters. Most often this is the interpreter, but in many of the countries it is also the government. Sometimes it also depends on the type of assignment, such as for conferences, these fees are in many cases determined by the interpreter.

Table 5.16 Who determines the fee for freelance interpreters?				
Country / region	**interpreter**	**government**	**employer**	**client**
Albania		x		
Austria	x	x		
Belgium - Flanders	x	x	x	x
Belgium - Wallonia	x	x	x	x
Bosnia-Herzegovina	x	x	x	x
Bulgaria	x	x		
Croatia	x			x
Cyprus		x		
Czech Republic	x			x
Denmark		x		
Engl. W&NI - ASLI	x			x
Engl. W&NI - NUBSLI	x	x	x	x
Estonia			x	
Finland[98]	x		x	x
France	x	x		x
Georgia			x	
Germany	x	x		x
Greece	x	x		
Hungary	x	x		x
Iceland	DK	DK	DK	DK
Ireland	x			
Italy - ANIMU	x			
Italy - ANIOS	x	x		x
Kosovo		x		
Latvia			x	
Lithuania	x			x
Luxembourg	x			
Malta[99]	x			
Moldova	x	x	x	x
Montenegro	x	x		
Netherlands[100]		x		
Norway		x		
Poland	x	x	x	x
Portugal	x		x	x
Romania	x			x
Russia	x		x	x
Scotland	x			
Serbia	x	x		x
Slovakia	x			
Slovenia		x		
Spain	x	x	x	x
Sweden			x	x

[98] Mainly the Social Insurance Institution of Finland (KELA) sets the prices through procurement.
[99] The interpreter association created fee guidelines.
[100] In case of commercial (non-governmental funded) assignments.

Table 5.16 Who determines the fee for freelance interpreters?				
Country / region	interpreter	government	employer	client
Switzerl.-Fr.[101]	x		x	
Switzerl.-Ge.	x			
Switzerl.-It.[102]				
Turkey	x		x	x
Ukraine[HI]	x	x	x	x

DK=Don't know

5.17 Reasonable pay

The participants were asked if the fees for interpreters were considered reasonable, the results can be found in Figure 5.21 and in a map in Figure 5.22. A total of twenty-two countries and regions (48%) indicate that they consider the payment for interpreting services reasonable, compared to 24% of the countries in 2020.

Fig. 5.21 Does an interpreter earn reasonable payment for the work they do? (N=46)

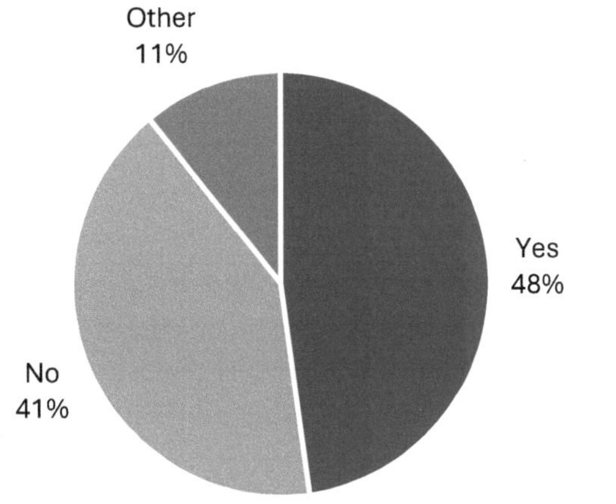

At the same time for twenty countries (41%) the pay is still not considered reasonable:

- Croatia: It is better than in previous years. However, given the current economic situation, it still leaves interpreters living slightly below the average citizen's standard of living.
- Engl. W&NI-NUBSLI: Depends on the individual interpreter.
- France: It depends on the region. Some interpreters feel they earn a good living, but most do not. Compared to spoken language interpreters, when working in the same settings sign language interpreters are paid less.
- Belgium-Flanders: The situation has improved over the last years, but compared to other countries, we are still behind. Especially considering the requirement of a master's degree for interpreters, the payment is not adequate.
- Ukraine[HI]: Most full-time or part-time interpreters who work for deaf associations or in the educational field are underpaid.

[101] Rates are set independently by freelance interpreters but are generally harmonized within the profession. Procom, as the main coordinating agency, also influences standard rates through its contracts and practices.
[102] Procom.

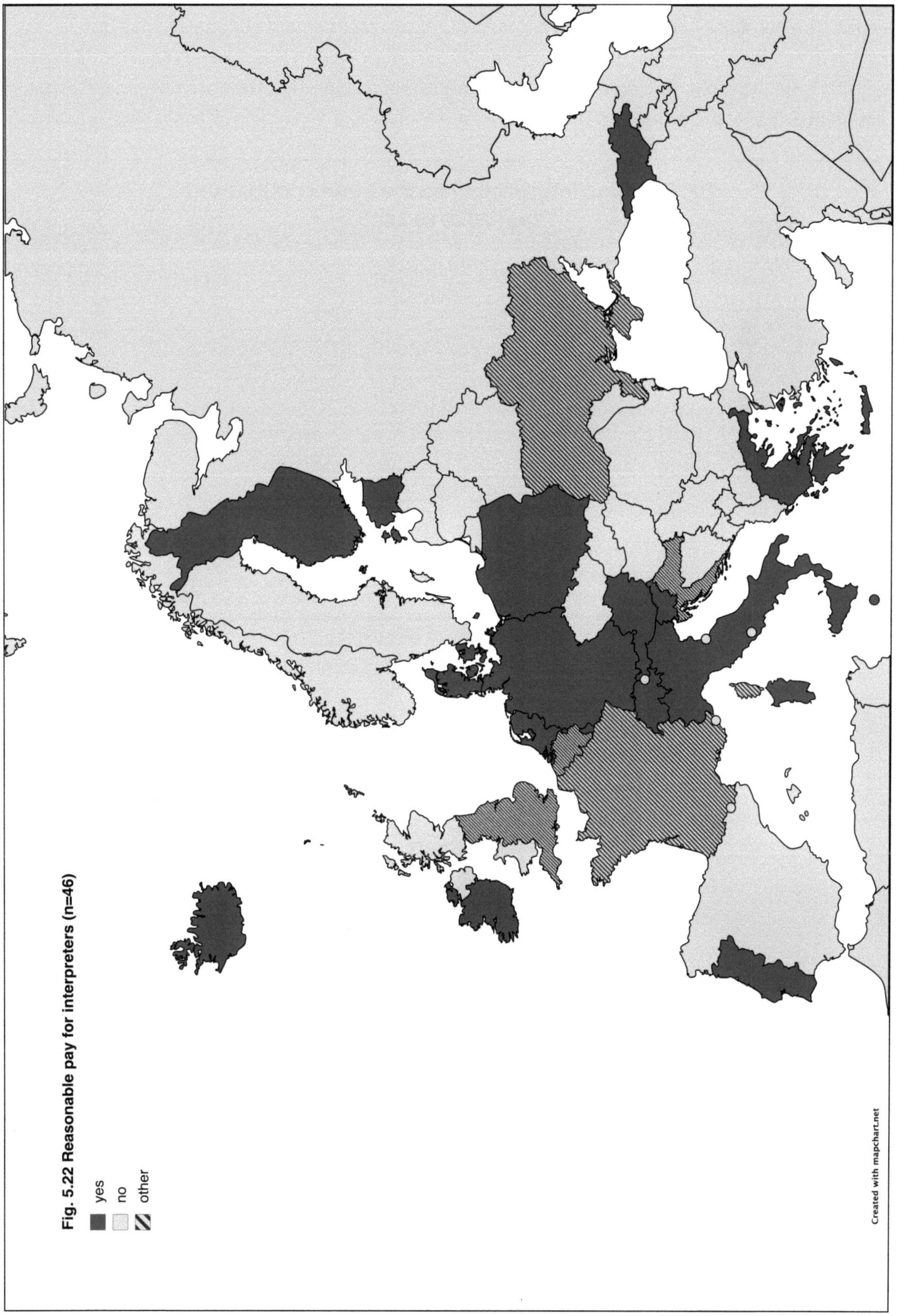

Fig. 5.22 Reasonable pay for interpreters (n=46)

5.18 Interpreting as a main occupation

Overall, the number of countries where interpreting can be the main occupation has not changed (Fig. 5.23; Fig. 5.24). In six countries this is not possible: Bosnia-Herzegovina, Bulgaria, Italy-ANIMU, Lithuania, Romania, Spain. Five countries indicate that they are not sure: Cyprus, Italy-ANIOS, Kosovo, Serbia, and Ukraine.

Fig. 5.23 Can interpreting be the main occupation? Year 2020 vs 2025

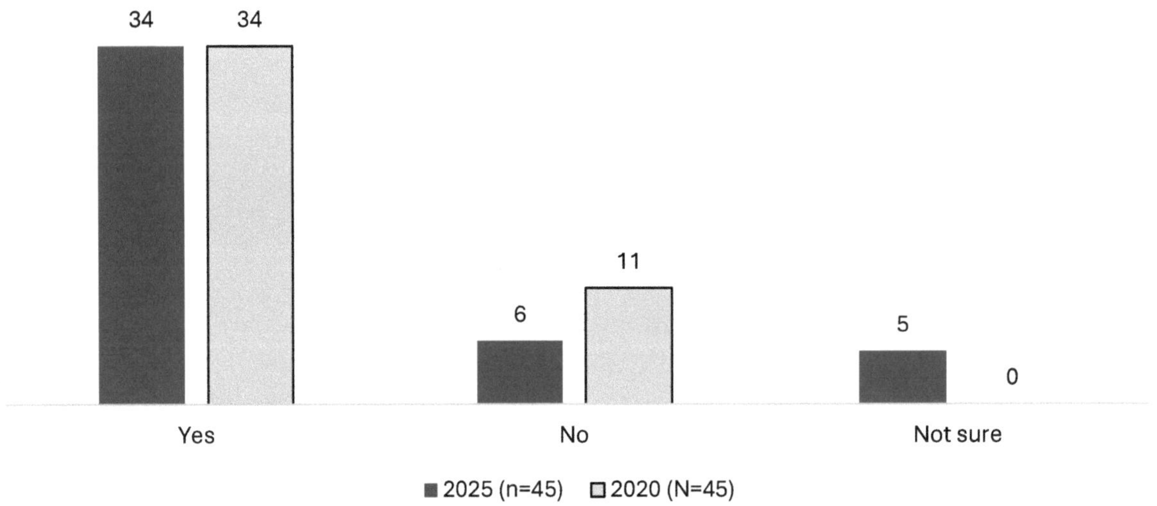

Fig. 5.24 Can interpreting be the main occupation, or is a second job needed to secure income (n=45)

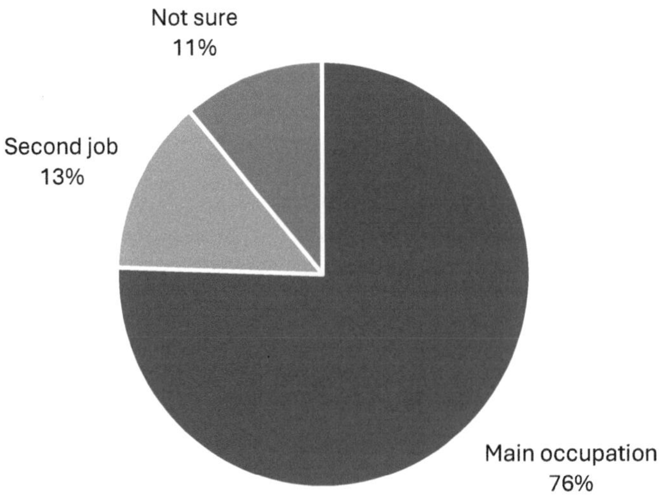

5.19 Budget limitations

As in 2016 and 2020, around twenty countries indicate that the government has partial budget limitations for the interpreting services in that country or region (Fig. 5.25 and Table 5.17). This means that access of deaf persons to interpreting services may be restricted because of budget constraints.

Fig. 5.25 Does the government have annual budget limitations on interpreting fees?

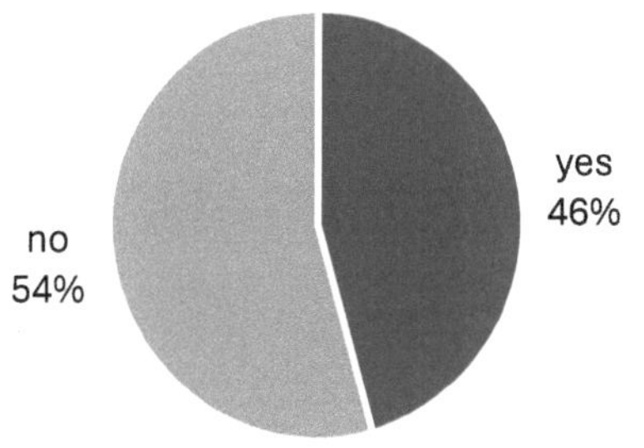

Table 5.17 Does the government have a total annual budget limitation for the interpreting services in your country/region?			
Country / region			
Albania	no	**Latvia**	yes
Austria	no	**Lithuania**	no
Belgium - Flanders	yes	**Luxembourg**	yes
Belgium - Wallonia	yes	**Malta**	yes
Bosnia-Herzegovina	no	**Moldova**	yes
Bulgaria	yes	**Montenegro**	no
Croatia	yes	**Netherlands**	yes
Cyprus	yes	**Norway**	no
Czech Republic	yes	**Poland**	no
Denmark	no	**Portugal**	no
Denmark[DI]	yes	**Romania**	no
Engl. W&NI	no	**Russia**	yes
Estonia	yes	**Scotland**	*No reply*
Finland	no	**Serbia**	no
France	no	**Slovakia**	yes
Georgia	yes	**Slovenia**	no
Germany	no	**Spain**	yes
Greece	yes	**Sweden**	no
Hungary	yes	**Switzerland-It.**	yes
Iceland	yes	**Switzerland-Fr.**	no
Ireland	no	**Switzerland-Ge.**	no
Italy - ANIMU	yes	**Turkey**	no
Italy - ANIOS	no	**Turkey**	no
Kosovo	no	**Ukraine**[HI]	no

5.20 Working in paid employment

In the survey the respondents were asked what an interpreter's average annual salary (in euro, excluding taxes) would be when working full-time for an agency or employer. The responses were so inconsistent that they have not been included in this publication.

5.21 Development of the profession

The sign language interpreter profession has been developing across Europe in various ways. In the following sections the respondents report on the development of the working conditions during the last five years, the impact of the Covid19 pandemic and AI, and where they see the main challenges and opportunities.

5.22 Working conditions – development

Looking back at the development of interpreters' working conditions during the last five years, about half of the respondents reports that in their country these have improved (Fig. 5.26). Table 5.18 details the explanations provided per country.

Fig. 5.26 Interpreter working conditions per country/region during the last five years (N=47)

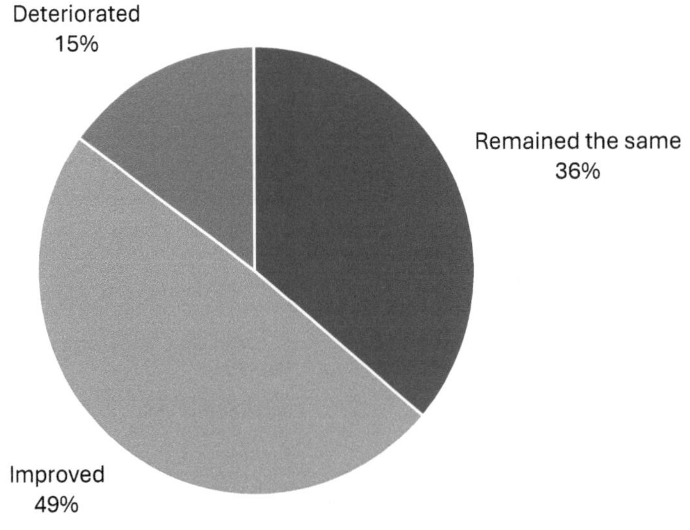

Table 5.18 Classification of the development of the working conditions for sign language interpreters during the last five years (2020-2025)

Improved, because:	
Austria	Increasing demand and certain conditions can be met.
Belgium - Flanders	Over the last years, the fee has been raised it is possible to work as a full-time interpreter, but a lot of interpreters combine interpreting with another job due to financial security.
Croatia	Salaries were increased because of laws passed last year.
Cyprus	There is greater need for interpreting, as more deaf individuals are involved in more activities, and the consequence is that more people are informed about interpretation and they respect the work of the interpreters as well as they recognize the importance of the provision of interpretation and the role of interpreters.
Czech Republic	NA
Denmark[DI]	Over the past four years, the working conditions for sign language interpreters in Denmark have generally improved. Prices have increased, tender conditions have become more favorable, and demand for interpreting services has grown. This has created better prospects for many interpreters, although challenges remain, such as uneven access to assignments and risks of overwork in high-demand areas. Overall, the development can be classified as a clear improvement.
Georgia	The legal framework has improved.
Germany	Due to lobbying.
Germany[DI]	Only very slight improvement and awareness among hearing colleagues and among clients. Media translation has long been the exclusive field of work for deaf interpreters. However, more deaf people who have no training or qualifications are being employed.
Greece	They have improved, as we have established the practice of working in pairs, many institutions now take care to provide us with good acoustic conditions, and we are treated on an equal footing with spoken language interpreters.
Iceland	More of teamwork, taking care of working hours.
Ireland	CISLI has worked hard on behalf of interpreters. There is greater collective and individual action by interpreters and the demand for interpreting services is high.
Italy - ANIOS	There are new fields of work, like remote jobs that give opportunities to interpreters to work full time.
Latvia	Raised salary, phones have been issued for work purposes, social guarantees.
Malta	More CPD, more Supervision, more opportunities for progression, improved collective agreement.
Netherlands	More online and deploying team interpreters more quickly is an improvement.
Poland	In recent years, working conditions for sign language interpreters in Poland have improved. It has become increasingly possible to work full time as an interpreter, particularly through institutional contracts and accessibility-related employment opportunities. The implementation of the Accessibility Act (2019) has also strengthened interpreters' position in negotiations with public institutions, as these institutions are now legally required to provide interpreting and ensure accessibility. This legal obligation has made interpreters' work more visible, stable, and recognized as an essential professional service.
Portugal	By gaining visibility with more online content being interpreted and television/media content being interpreted, this visibility also brought more social recognition on the importance of having accessibility for the Deaf, so new job opportunities were created from this development.
Russia	Can interpreting be the main occupation - yes, for the central region, most of the interpreters in remote regions work part-time. Improved because Inclusive institutions in Moscow and St. Petersburg have paved the way for the evolvement of the profession.
Scotland	Post covid interpreters now have more flexibility with working onsite or remotely.
Slovakia	Improved because of the lobby from deaf association but only in one district (Trnava district) where the employer is deaf and is at the same time chair of the deaf association.
Switzerland-Fr.	The collective labor agreement (CCT) has evolved; we now benefit from inflation adjustment and better cancellation conditions. Procom employees were able to take an active role in the negotiations and managed to influence the outcome in their favor.
Switzerland-Ge.	The collective agreement improved for the interpreters; we now always get paid the hours we are hired for (before we got paid the hours we actually worked).

Table 5.18 Classification of the development of the working conditions for sign language interpreters during the last five years (2020-2025)			
Turkey	In recent years, awareness of sign language interpreting in Türkiye has increased. Public institutions, particularly in the fields of justice and healthcare have begun to employ more interpreters. In addition, access to interpreting services has become easier through remote video interpreting systems and digital platforms. These developments have positively affected both the working conditions and the visibility of the profession.		
Ukraine[HI]	As demand for sign language interpreting services at conferences, on TV, in educational and awareness-raising video courses, etc. has gone up, some interpreters can make a decent living. But for most, interpreting isn't a money-maker.		
Deteriorated because			
Denmark	Salary has not gone up significantly (enough).		
Finland	The estimate is that the working conditions might have been deteriorated slightly. The Social Insurance Institution of Finland (KELA) sets the prices through its procurement process and also effects other working conditions with its guidelines on how the service is provided.		
France	As for the working conditions per se, they did not change a lot. But financial resources have deceased a lot, mostly since the last government's turnover.		
Hungary	Depreciation: the (freelance) salary Is the same since 2009.		
Montenegro	We do not have sign language rights regulated by law.		
Spain	For many years, the profession has been characterized by unstable employment. Furthermore, in recent years, when private entities (including those representing the deaf community) simply sought "replacements" from younger generations willing to endure this job insecurity, most interpreters have left the profession. Today, the fact that anyone—with or without official qualifications—can perform our work does not contribute to respect for the profession, which exacerbates the fact that fewer people want to dedicate themselves to interpreting, even though demand has increased. To make matters worse, our salaries are getting lower. It is true that public institutions such as the Álava Provincial Council have recognized the "grade" category for our positions in ILS (Collective Agreement for Centers and Services for People with Disabilities in Álava), which has resulted in a salary increase. However, organizations representing deaf people have positioned themselves against this measure, even though they are not the ones who directly bear the cost of the salary.		
Sweden	More procurements with wildly different terms make it harder to navigate and the salary does not match the inflation and rise in the cost of living.		
Remained the same			
Albania	Italy - ANIMU	Moldova	Slovenia
Belgium - Wallonia	Engl. W&NI-NUBSLI	Lithuania	Romania
Bosnia-Herzegovina	Estonia	Luxembourg	Serbia
Bulgaria	Kosovo	Norway	Switzerland - It.
Engl. W&NI-ASLI			

5.23 Impact of the Covid19 pandemic on the profession

The Covid19 pandemic had an impact on the signed language interpreter profession. The responses show four main themes (see Fig. 5.27):

1. Visibility of the profession increased, especially through TV, (social media), and public briefings such as press conferences.
2. Negative effects also appeared: deteriorated working conditions, reduced in-person work, income loss, isolation, and access barriers.
3. Some countries reported structural gains, such as more formal acceptance of remote interpreting or better accessibility planning.
4. A few reported little or no overall impact.

Table 5.19 shows the countries and which main impact they report. Four countries did not answer this question: Finland, Greece, Hungary, and Kosovo. The pandemic seems to have been a turning point toward remote interpreting, the dominant cross-country pattern. At the same time, respondents describe a mixed legacy: more recognition and accessibility on one hand, but more precarious working conditions and uneven access on the other.

Fig. 5.27 Impact of Covid19 pandemic on the profession (n=43)

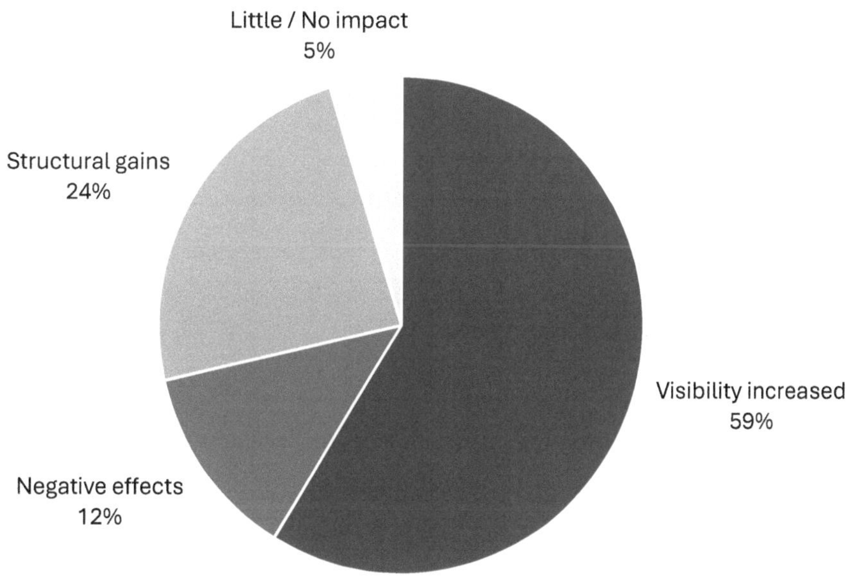

Table 5.19 Impact of Covid19 pandemic on the sign language interpreter profession (n=43)				
Country / region	**Visibility Increased**	**Negative Effects**	**Structural Gains**	**Little / No Impact**
Albania				X
Austria	X		X	
Belgium - Flanders	X		X	
Belgium - Wallonia	X		X	
Bosnia and Herzegovina	X			
Bulgaria				X
Croatia	X			
Cyprus	X		X	
Czech Republic	X			
Denmark	X		X	
Engl. W&NI-ASLI	X			
Engl. W&NI-NUBSLI	X	X		
Estonia	X			
France	X			
Georgia	X			
Germany	X		X	
Iceland				X
Ireland	X			
Italy - ANIMU	X			
Italy - ANIOS	X			
Latvia	X		X	
Lithuania	X			
Luxembourg	X		X	
Malta		X		
Moldova	X			
Montenegro		X		
Netherlands	X			
Norway	X	X		
Poland	X		X	
Portugal	X		X	
Romania	X			
Russia	X			
Scotland	X			
Serbia	X		X	
Slovakia	X		X	
Slovenia	X		X	
Spain	X	X		
Sweden	X	X		
Switzerland – Fr.	X		X	
Switzerland – Ge.	X			
Switzerland – It.		X		
Turkey	X	X	X	
Ukraine[HI]	X			

5.24 Impact of AI on the profession

For the first time in this survey the countries were asked if AI has had an impact on the sign language interpreting profession. From the responses six different themes were identified (Fig. 5.28 and Table 5.20):

1. No / minimal impact (67%): Most countries report no current impact of AI on the SLI profession.
2. Research & development phase (11%): Ongoing experiments (avatars, recognition systems), but not usable in practice yet.
3. AI as supportive tool (limited) (7%): Minor uses (e.g., preparation, written communication) but not for core interpreting.
4. Perceived threat / concern (7%): A few express fears of replacement.
5. AI as future potential / emerging topic (4%): Many responses frame AI as something coming, not yet present.
6. Minor practical effects (4%): Very limited real-world influence so far.

Fig.5.28 Impact of AI on the profession (n=43)

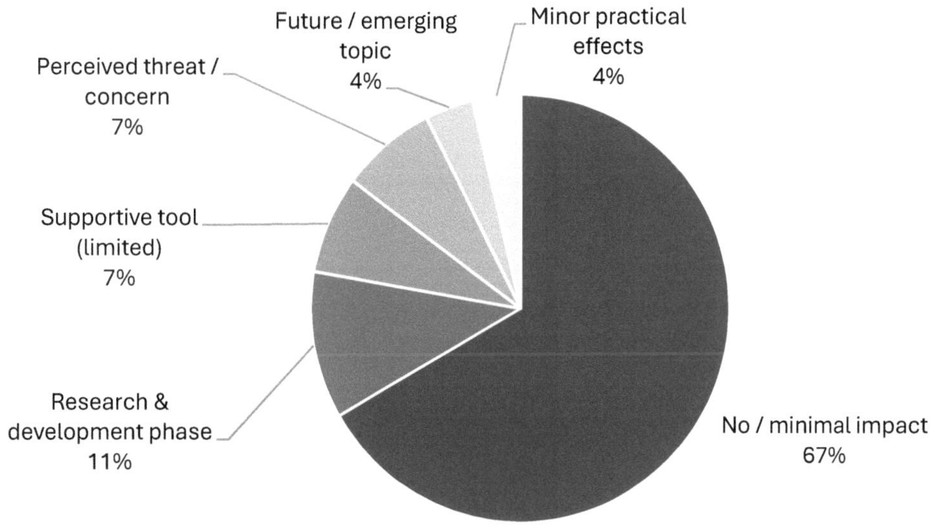

Table 5.20 AI impact on sign language interpreter profession (n=43)

Country / region	No / minimal impact	Future potential / emerging topic	Research & development phase	Supportive tool (limited)	Perceived threat / concern	Minor practical effects
Albania	X					
Austria	X					
Belgium-Flanders	X				X	
Belgium-Wallonia	X					
Bosnia and Herzegovina	X					
Bulgaria	X					
Croatia	X					
Cyprus	X					
Czech Republic	X					
Denmark	X					
Engl. W&NI-ASLI					X	
Engl. W&NI-NUBSLI					X	
Estonia				X		X
Finland	X					
France	X				X	
Germany	X					
Iceland	X					
Ireland	X		X			
Italy-ANIMU	X					
Kosovo	X					
Latvia	X					
Lithuania	X					
Luxembourg	X					
Malta	X					
Moldova	X					
Montenegro	X					
Netherlands	X					
Norway	X					
Poland	X		X			
Portugal	X					
Romania	X					
Russia			X			
Scotland	X					
Serbia	X			X		X
Slovakia	X					
Slovenia	X					
Spain	X					
Sweden	X					
Switzerland-Fr.		X	X	X		

Table 5.20 AI impact on sign language interpreter profession (n=43)						
Country / region	No / minimal impact	Future potential / emerging topic	Research & development phase	Supportive tool (limited)	Perceived threat / concern	Minor practical effects
Switzerland-Ge.			X			
Switzerland-It.	X					
Turkey		X	X	X		
Ukraine[HI]	X					

5.25 Main professional opportunities and challenges in 2025

The countries reported on their current (2025) main professional opportunities and challenges (Fig. 5.29; Table 5.21). Across countries, in the responses the main themes are:

Challenges:

1. Shortage of interpreters
2. Inadequate pay and lack of funding
3. Training and development gaps
4. Insufficient regulations

Opportunities:

5. Support by associations
6. Growing demand
7. Collaboration

Fig. 5.29 Challenges and opportunities in the profession in the countries in 2025 (n=43)

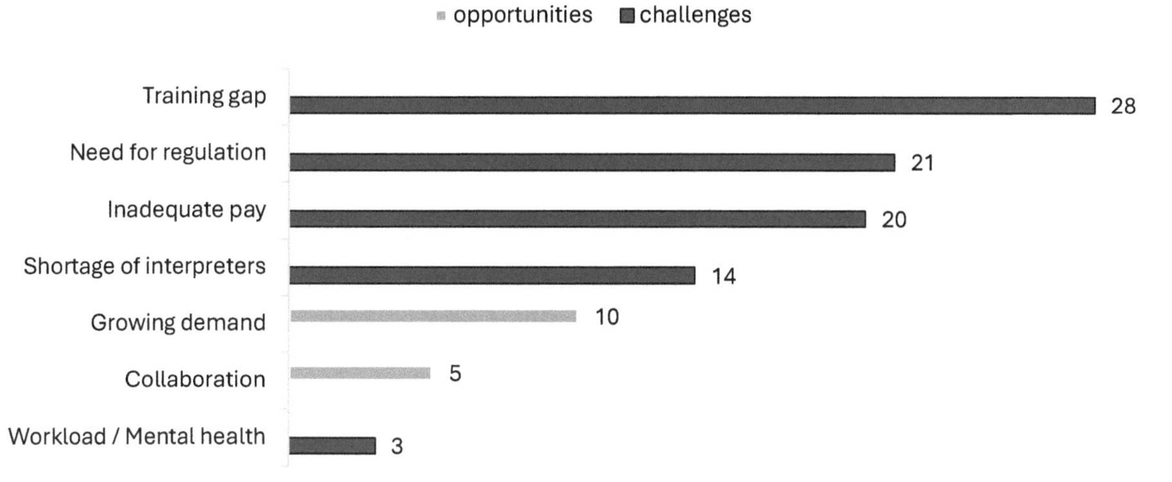

1. Shortage

The central tension is the combination of a high shortage of interpreters and a continuous high demand:

- Many respondents mention a shortage of interpreters
- At the same time, there is growing demand for interpreting services

This creates a key paradox: demand is increasing, but systems cannot supply enough qualified professionals.

2. Inadequate pay

In addition, precarious working conditions are widespread:

- Poor pay and unstable funding
- Often linked to:
  - Freelance work
  - Lack of governmental support
  - Project-based employment

It shows that the profession is not yet economically sustainable in many regions.

3. Training gaps

Training and certification gaps are critical, which is a continuous theme, indicating that there is a lack of:

- Formal education programs
- Certification systems
- Continuous professional development

These training gaps connect directly to:

- Quality concerns
- Lack of regulation
- Uneven professional standards

4. Insufficient regulations

Countries highlight:

- Absence of formal regulations
- Lack of legal recognition: laws exist but are not implemented
- Weak or absence of accreditation systems

This weakens job stability, service quality, and professional development.

In addition, there are some concerns around well-being of interpreters, such as:

- Workload / burnout / mental health appears
- But less frequently than structural issues

This seems more of an emerging issue, especially in high-demand contexts.

5. Support by associations

Where professional associations and institutions are active, they create important opportunities and improved access to interpreting work, such as:

- Supportive working networks
- Collaboration with deaf associations
- National initiatives

These efforts can strengthen professional visibility, standards, and long-term career sustainability.

Countries mention opportunities; however, they appear to be secondary.

6. Growing demand

Sign language interpreting services are more frequently requested, especially to provide accessibility:

- Growing demand (public sector, media, accessibility)
- Some increased visibility of the profession

7. Collaboration

Collaboration is emerging but not very present. Some references are made to:

- Collaboration with deaf interpreters
- Professional networks
- Agencies

However, these developments are considered less prominent, which indicates these are not yet structurally embedded.

In conclusion, the findings reveal that the signed language interpreting profession across countries is characterized by a combination of high demand and structural fragility. While opportunities are increasing—particularly through expanding accessibility needs—these are overshadowed by persistent systemic challenges, including workforce shortages, inadequate pay, limited training pathways, and insufficient regulation. The profession remains in a phase of ongoing professionalization, with uneven institutional support and limited standardization across contexts. Overall, the data suggest that growth in demand is not yet matched by sustainable professional infrastructures.

Table 5.21 Main professional challenges and opportunities in 2025

Country / region	Challenges					Opportunities		
	Short-age	Inadequate pay / Lack of funding	Training gap	Workload / Mental health	Insufficient regulation	Support by associations	Growing demand	Colla-boration
Albania	X	X			X			
Austria		X	X		X			
Belgium-Flanders	X	X	X	X		X		X
Belgium-Wallonia	X	X						
Bosnia and Herzegovina	X	X	X		X		X	
Bulgaria								
Croatia			X			X		
Cyprus	X	X	X		X			X
Czech Republic	X		X		X			
Denmark				X			X	
Engl. W&NI-ASLI								
Engl. W&NI-NUBSLI		X	X		X			
Estonia	X			X	X		X	
Finland								
France		X	X		X	X		
Georgia		X	X					
Germany			X					
Iceland		X						
Ireland	X						X	
Italy-ANIMU		X						
Italy-ANIOS			X		X			
Kosovo	X	X						
Latvia			X					
Lithuania		X	X		X			
Luxembourg			X		X			
Malta			X					X
Moldova		X	X			X	X	
Montenegro					X			
Netherlands			X		X			
Norway								
Poland			X		X		X	
Portugal			X			X	X	
Romania	X	X	X				X	
Russia		X	X					
Scotland								
Serbia	X		X		X		X	
Slovakia	X	X	X		X			

Table 5.21 Main professional challenges and opportunities in 2025								
Country / region	Challenges					Opportunities		
	Short-age	Inadequate pay / Lack of funding	Training gap	Workload / Mental health	Insufficient regulation	Support by associations	Growing demand	Colla-boration
Slovenia			X		X			
Spain		X	X		X	X		
Sweden						X		X
Switzerland-Fr.		X	X		X			X
Switzerland-Ge.					X		X	
Switzerland-It.	X							
Turkey	X	X	X		X		X	
Ukraine[HI]			X	X				

Appendices

1. Survey
2. Overview of facts by country
3. Overview of all figures

See also https://www.mayadewit.nl/european-study for:

- Links to national complaint mechanisms
- Links to national registration or accreditation bodies

Appendix 1 – Survey sign language interpreting in Europe

Sign Language Interpreting in Europe - 2025

Your contact details
Please fill out your name and email. I can then contact you if I need a clarification on your answers.

Your country/region

1. Which country or region are you reporting on?
2. How many national or regional associations for sign language interpreters does your country have?
3. Is the presence of more than one association beneficial for deaf persons and interpreters?
   [ ] Yes, because [ ] No, because
4. Are you reporting on behalf of your national/regional association? [ ] yes [ ] no

NOTE: By answering this survey, you confirm that you are representing your national or regional association. If you are not responding on behalf of your national association, we assume that you respond on behalf of your country/region.

Your association of sign language interpreters

The following questions are specifically about your association and its members.

5. What is the official name of your association (in the original language)?
6. What is the acronym of your association (in the original language)?
7. What is the email address of your association?
8. What is the web address of your association?
9. In which year was your association officially established?
10. How is your association funded? (tick all that apply)
    - Membership fees
    - Donations
    - Grants
    - Government funding
    - Other __________
11. What does your association have? (tick all that apply and if these are volunteer or paid positions)

	Volunteer	Paid	NA
Board	❑	❑	❑
President/chair of the association	❑	❑	❑
Office staff	❑	❑	❑
Daily manager/executive director/CEO	❑	❑	❑
Chief financial officer	❑	❑	❑
Advisory board	❑	❑	❑
Ethical/disciplinary/complaints committee	❑	❑	❑
Working groups and/or committees	❑	❑	❑
Other	❑	❑	❑

Association membership

12. Does your association have regional associations as members? [ ] yes [ ] no

In the following question an 'interpreter' is referred to as someone who has a qualification or some official recognition as a sign language interpreter.

13. What percentage of working interpreters in your country/region is a member of your association? (if you do not know the exact percentage, please give an estimate)
14. How many individual interpreter members does your association have or represent? Note: if you are an umbrella organization with regional associations as members, please state the total number of members represented by all the regional associations that are part of your organization.
15. How many individual interpreter members identify as hearing, deaf or hard of hearing?

- Hearing __________
- Deaf __________
- Hard of hearing __________

16. Does your association collect members information on gender, race and ethnicity by asking members to self-identify? [ ] yes [ ] no, because [ ] no, it is illegal in our country/region to collect such data

Note: If your association does not track this information, if possible, please provide your best estimate and indicate that the data is estimated rather than based on formal records. These questions are optional.

17. Do you know the (estimated) percentage of gender distribution among the members in your association? [ ] I can give an (estimated) percentage [ ] I do not know [ ] I prefer not to answer
18. What is the (estimated) percentage of gender distribution among the members in your association? [ ] Female __________ [ ] Male __________ [ ] Non-binary __________ [ ] Other __________
19. Are ethnic and racial minorities from your country/region represented in your association? [ ] yes [ ] no
20. Which requirements must individual interpreters meet to become a member of your association? (tick all that apply)
    - Competency assessment or admission exam conducted by a committee
    - Degree from a national sign language interpreter training program
    - Minimum educational degree from a college or university (not necessarily in interpreting)
    - Agree to work in accordance with the national interpreter code of ethics
    - Experience: Such as a minimum number of hours/days/years working as an interpreter
    - Recommendation letter from individual interpreters
    - Recommendation letter or certificate from the national deaf association
    - Recommendation by a current member of the association
    - Citizen of the country
    - No criminal record
    - Other
21. What are the benefits for sign language interpreters as a member of your association? (tick all that apply)
    - Representation of the profession
    - Guidance and advice
    - Current information via email
    - Newsletters
    - Discounts on training, conferences and materials
    - Members only section on website
    - Other, namely
22. Are the requirements to become a member of your association the same for deaf and hearing interpreters? [ ] yes [ ] no [ ] NA
23. Does your association have sign language interpreters as members who interpret for persons who are deafblind? [ ] yes [ ] no
24. How many non-interpreter members in total does your association have? (these can be individuals and associations)
25. What is the annual membership fee for an individual interpreter member (in euros)? Note: If your membership fee is not in euro, please go here to the currency converter: www.oanda.com/currency/converter/

Association's aims and quality control

26. Does your association focus on quality control of sign language interpreters? [ ] yes [ ] no
27. What are the main aims of your association? (tick all that apply)
    - Offer training and continuing education
    - Ensure quality control
    - Ensure good working conditions
    - Negotiate interpreting fees
    - Provide a network of support
    - Cooperate with stakeholders, e.g. national deaf association
    - Other, namely
28. If your association has other members besides individual interpreter members, please select all that apply
    - Student or trainee interpreters

- Trainers and educators
- Institutes
- Associations
- Companies
- Other, namely

29. Does your association collaborate with the national/regional deaf organization? [ ] yes [ ] no
30. What are the main issues your association is currently facing?

If you have any comments on your association or the association's aims and quality control, please note them here.

Accreditation or registration of interpreters

Some countries/regions have a special institute or body which is responsible for the official accreditation or registration of sign language interpreters: who is an interpreter and who is not. This body is not the same as the institute issuing an educational diploma. The body can be independent or can be part of another association.

31. Is there an official body which officially registers/accredits sign language interpreters in your country/region? [ ] yes [ ] no
32. How many bodies are present in your country that register or accredit signed language interpreters?
33. Is having several bodies for accreditation and registration beneficial to interpreters and deaf persons? [ ] yes [ ] no
34. What is / are the names of these official accreditation/registration bodies?
35. What is the official name of the main registry/accreditation body (in your language)?
36. If the main registration/accreditation body has a website, please fill out the web address here:
37. Is the main registration/accreditation body part of your national association of sign language interpreters?* [ ] yes [ ] no
38. Is the main registration/accreditation body part of another association? [ ] yes [ ] no
39. What are the prerequisites to be in the main registry/accreditation body as a sign language interpreter?
40. Is it compulsory for sign language interpreters to be registered/accredited by this main body to work as an interpreter?* [ ] yes [ ] no [ ]other
41. Do you know the percentage of gender representation of the sign language interpreters in this main registration/accreditation body? [ ] yes [ ] no
42. What is the percentage of gender representation of the sign language interpreters in this main registration/accreditation body?
    - Female __________
    - Male __________
    - Non-binary __________
    - Other __________
43. How many sign language interpreters in total are registered/accredited by the main body?
44. Can the following sign language interpreters register in the main registration/accreditation body? (tick all that apply)
    - Hearing interpreters
    - Deaf and hard of hearing interpreters
45. How many hearing sign language interpreters are currently registered/accredited?
    - I have an exact number
    - I have an estimated number
    - I do not know
46. How many deaf sign language interpreters are currently registered/accredited?
    - I have an exact number
    - I have an estimated number
    - I do not know
47. How many interpreters are registered/accredited who interpret for deafblind persons?
    - I have an exact number
    - I have an estimated number
    - I do not know

If you have any comments on the topic of registration/accreditation, please note them here:

Deaf interpreters' developments

All questions in this survey are for all interpreters, deaf and hearing. However, as deaf interpreters in Europe generally do not have the same opportunities as hearing interpreters, this section asks specifically about the development of the profession for deaf interpreters in your country/region. Included is an additional question about deaf translation work.

48. What term is mostly used in your country/region when referring to deaf sign language interpreters?
    - Deaf interpreter
    - Deaf sign language interpreter
    - Sign language interpreter
    - Interpreter
    - Other
49. How do hearing sign language interpreters support the development and/or the work of deaf interpreters in your country/region?

Employment and work of deaf interpreters

50. How many deaf interpreters work as interpreters in your country/region?
    - I know the exact number
    - I have an estimated number
    - There are no deaf interpreters working in my country/region
    - I do not know
51. Where are deaf interpreters recruited to work mostly? (tick the 3 most common)
    - News or media broadcasts
    - International or multilingual settings
    - Legal settings (e.g. court, police, immigration services)
    - Interpreting for deafblind persons
    - Conference type settings
    - Translation services - not live interpreting (e.g. pre-recorded informational posts)
    - Other
52. What are the main barriers for deaf interpreters in your country/region to be recruited as an interpreter?
53. How do deaf interpreters in your country/regions work? (tick all that apply)
    - As a co-interpreting team, with the feed/relay of hearing sign language interpreter
    - Via live captions / subtitles
    - I do not know
    - Other
54. In case of co-interpreting teams (tick all that apply)
    - The deaf interpreter can use their personal right to an interpreter to finance the fees of the hearing interpreter
    - The deaf interpreters can decide who they want to work with as co-interpreter
    - Other __________
55. How is the payment for the interpreting services by deaf interpreters organized?
    - The same as for hearing interpreters
    - There is a specific payment rule
    - There is no established funding available for interpreting services by deaf interpreters
    - I do not know
56. How are the assignments distributed for deaf persons working as interpreters and/or translators?
    - All work is in interpretation
    - All work is in translation
    - Most work is in translation, some in interpretation
    - Most work is in interpretation, some in translation
    - I do not know

    Comments on the work of deaf interpreters

Training of deaf interpreters

57. How are deaf interpreters trained in your country/region?
    - By attending any sign language interpreting program
    - Through a dedicated deaf interpreter program

- In practice, there is no training available
- Other
- I do not know

58. How many trained deaf interpreters are there in your country/region?
    - I know the exact number
    - I have an estimated number
    - I do not know
59. Are deaf persons trained to interpret via live captions and/or as co-interpreter with a hearing interpreter? (tick all that apply)
    - Via live captions/subtitles
    - As a co-interpreting team, with the feed/relay of hearing sign language interpreter
    - Other
60. In general, rate the following for deaf interpreters in your country/region

	1	2	3	4	5
How would a deaf interpreter classify the available interpreter training for deaf interpreters?	❑	❑	❑	❑	❑
How is the access to mentoring, and professional development opportunities for deaf interpreters?	❑	❑	❑	❑	❑

61. What are the main issues in training deaf interpreters in your country/region? Note: more detailed questions on the training of deaf and hearing sign language interpreters are in the upcoming section on Education.

Comments on training deaf interpreters:

New developments

62. How has the profession of sign language interpreters developed for deaf interpreters in your country/region during the last 5 years?
    - It has remained the same
    - It has improved
    - It has deteriorated
63. What changes have you seen in the opportunities for deaf interpreters in your country/region?
64. What are the biggest barriers deaf interpreters currently face in your country/region?
65. What are the most critical measures needed to strengthen the professional role and visibility of deaf interpreters nationally or regionally?
66. Are deaf interpreters included in official standards, qualification frameworks, or professional guidelines in the interpreting sector?
    - Yes
    - Partially
    - No
67. Can you give examples of how deaf interpreters are included in official standards, qualification frameworks, or professional guidelines in the interpreting sector?
68. Why are deaf interpreters not included in official standards, qualification frameworks, or professional guidelines in the interpreting sector?

This was the last question on the topic of deaf interpreters. If you have any comments on the topic, please note them here:

Complaint mechanism

69. Does your country/region have a complaint mechanism for clients of sign language interpreting services?
    - Yes
    - No
70. Is there another way the interpreter receives feedback from clients?
    - Yes
    - No

71. Who can file a complaint? (tick all that apply)
    - Deaf/deafblind clients
    - Hearing clients
    - Interpreters (reporting on their colleague)
    - Other
72. If the complaint procedure has a website, please provide it here:
73. Are deaf clients aware of this complaint procedure?
    - Most of them
    - Some of them
    - No
    - I do not know
74. When deaf clients are not happy with the services received, do they use the possibility to file a complaint?
    - Most of the time
    - Often
    - Sometimes
    - Rarely
    - Never
    - I do not know
75. Are hearing clients aware of this complaint procedure?
    - Most of them
    - Some of them
    - No
    - I do not know
76. When hearing clients are not happy with the services received, do they use the possibility to file a complaint?
    - Most of the time
    - Often
    - Sometimes
    - Rarely
    - Never
    - I do not know
77. What are the consequences for the interpreter when a complaint is proven valid?

If you have any further comments about this topic, please note them here

Education & training

78. Does your country / region have an educational program for sign language interpreters?
    - Yes
    - No
79. In general, can deaf persons in your country/region attend the same training program as hearing interpreters?
    - Yes
    - No
80. Are there (also) other possibilities for deaf persons to be trained as a sign language interpreter?

Educational programs for sign language interpreters

81. Do you need to have an interpreting degree or complete an interpreter education to legally work as an interpreter in your country/region?
    - Yes
    - No
82. How many different formal educational programs for sign language interpreters are there in your country or region? Note: a formal educational program takes place within an organized educational institution, such as a school, college, or university. It is characterized by a set curriculum, defined learning objectives, standardized assessment methods, and the awarding of recognized qualifications or degrees upon successful completion of the program.
    - 1
    - 2
    - 3

   - 4
   - 5 or more

   Do you have any comments in general about the education of sign language interpreters, please note them here.

83. *First educational program* (repeated for as many programs there are)
    - Name of the educational program
    - Name of the educational institution
    - City
    - Contact person
    - Email
    - Website
    - Comments
84. Program is offered:
    - Full time
    - Part time
85. Duration of the program (in months or years, please specify):
86. Number of ECTS (if applicable):
87. Does the program use the CEFR as reference for their national sign language?
    - Yes
    - No
    - I do not know
88. Is the program public or private?
    - Public
    - Private
89. What is the status of the program? (tick all that apply)
    - Officially supervised & recognized by the government
    - Recognized or supported by the sign language interpreter association
    - Recognized or supported by the national deaf association
    - Other
90. What is the total number of students currently enrolled?
    - I know the exact number of students
    - I have an estimate on the number of students
    - I do not know
91. What is the educational level of the program?
    - Vocational / folk high school
    - Bachelor at university of applied sciences
    - Bachelor at university
    - Master at university of applied sciences
    - Master at University
    - Post-graduate degree
    - Other
92. Does the program offer any minors?
    - Yes
    - No
93. Does the program also educate students to interpret from and to other languages? (tick all that apply)
    - English (as a foreign language)
    - International Sign
    - Other spoken languages
    - Other signed languages
    - I do not know
    - None of the above

94. Does the program educate students in:

	Yes	No	I do not know
Remote/Distance interpreting	❑	❑	❑
Team interpreting	❑	❑	❑
Co-interpreting (deaf and hearing interpreters in a team)	❑	❑	❑
Interpreting for deafblind persons	❑	❑	❑
Working with AI tools	❑	❑	❑

95. Does the program include an internship / traineeship?
   - Yes
   - No
96. What are the prerequisites to attend this educational program? (tick all that apply)
   - Pass an entry test/exam
   - A minimum level of a spoken language fluency
   - A minimum level of sign language fluency
   - High school diploma
   - Other
   - None
97. Can deaf and hearing students attend this program?
   - Yes
   - No
   - I do not know

   Comments:
98. Are there any specific or different requirements for deaf students to enroll in this program?
   - Yes
   - No
   - I do not know
99. Has the curriculum been adapted to educate deaf and hearing students in the same program?
   - Yes
   - No
   - I do not know
100. Are there any specific modules for deaf or hearing students only?
   - Yes
   - No
   - I do not know

No education

101. Your country/region does not have an interpreter training program. Where or how do the interpreters receive training?

Specific & continuing education

102. Is continuing education or further training required to remain certified or accredited as an interpreter?
   - Yes
   - No

   Comments:
103. What are the requirements for the continuing education or further training?
104. Can the interpreter specialize in a specific field (e.g. interpreting, legal, conference, etc.)?
   - Yes
   - No

105. Which specializations and training are available and/or formally recognized? (tick all that apply)

	Available	Formally recognized	N/A
Deafblind interpreting	❑	❑	❑
Legal	❑	❑	❑
Conference	❑	❑	❑
Theatre	❑	❑	❑
TV / media	❑	❑	❑
Medical	❑	❑	❑
Higher education	❑	❑	❑
Team/co-interpreting	❑	❑	❑
Other	❑	❑	❑

106. Do any of these specializations have specific requirements?
- Yes
- No

Any comments on continuing education or specializations, please note them here:

Mentoring / supervision

107. Does your country/region have a mentorship or supervisions program for sign language interpreters and/or interpreter student/trainees available?
- Yes
- No

108. Is the mentorship / supervision program: (tick all that apply)
- On a voluntary basis
- Formally recognized
- Mandatory

109. What type of mentorship / supervision program is it? (tick all that apply)
- For interpreting students
- For recent graduates
- Peer-to-peer
- For a specific setting or specialty
- Other

110. Who organizes the mentoring / supervision program?
- Interpreters themselves
- The interpreter training program
- Interpreter agencies
- Deaf / deafblind associations
- Other

Comments on mentoring and supervision:

Quality control

111. If there are any (other) specific measures in place in your country or region to ensure quality control of sign language interpreters, please state them here.

Profession regulated

112. Does the national legislation provide a definition of what a sign language interpreter is?
   - Yes
   - No
113. Is the sign language interpreter profession regulated and/or protected in your country? See here a more detailed explanation
   - Yes, for reserves of activities
   - Yes, protected title (without reserves of activities)
   - Yes, reserves of activities and protected title
   - Other
   - No
   - I do not know
114. Are there other professionals who are used instead of qualified interpreters? (for example, a teaching assistant in the classroom)
   - Yes
   - No

   If you have any comments on regulating the profession, please note them here

Sign Language

115. Approximately how many deaf and hard of hearing signing persons are there in your country/region?
116. Is your national/regional sign language(s) recognized by law or regulation in your country/region?
   - Yes
   - No
117. In which year was sign language officially recognized?
118. What is / are the names of the sign language(s) in your country/region? (in the original language)
119. Is the right to a sign language interpreter or interpreting services recognized by your country/region laws or regulations?
   - Yes
   - No
120. What do these rights to an interpreter consist of?
121. As of what age does a deaf/deafblind child/person have the right to an interpreter in an educational setting?
122. Who pushes for action to ensure the use of the legal right to an interpreter? (select the option that is most frequently the case)
   - Government
   - Individual deaf person
   - Deaf organization
   - Individual interpreter
   - Interpreter organization
   - Other
123. If you have an internet link to the legal text of the interpreting law in your country, please provide it here
124. How can a deaf person arrange an interpreting service? (tick all that apply)
   - Directly contacting the interpreter
   - Online database of interpreters
   - Through an interpreter referral agency
   - Through the deaf person's employer
   - Personnel responsible for interpreting services in the specific setting (hospital, university, etc.)
   - At the deaf association
   - At the interpreter association
   - Other
125. When an interpreting service is requested can the deaf or deafblind person choose or request a preferred interpreter?

	Yes, always	Yes, in certain settings	No
Deaf/hard of hearing person	❑	❑	❑
Deafblind person	❑	❑	❑

Employment

126. How many sign language interpreters are currently working in the country/region you are reporting on?
127. How many sign language interpreters who interpret for deafblind persons are currently available in your country/region?
128. Are there in general sufficient interpreters to meet the demand?
    - Yes
    - No
    - Not sure
129. If there is a lack of interpreters, where is this lack noticeable (tick all that apply):
    - In remote regions
    - In regions without an interpreter training program
    - In settings that demand additional interpreter competencies (for example in higher education or use of English as a foreign language)
    - In last minute assignments, such as a medical emergency, funeral, job interview, etc.
    - Other
130. How do interpreters work in your country / region? Please note: an agency or an employer can be private companies, deaf associations, public institutions, etc.
    - All as freelancers
    - All for an agency or employer
      Some as freelancers, some for an agency/employer
    - Mostly as freelancers, some for an agency/employer
    - Mostly for an agency/employer, some as freelancers
    - Other
131. Do interpreters in your country/region work:
    - Mostly onsite
    - Some onsite and some remote
    - Mostly remote
    - I do not know
132. Since the Covid-19 pandemic requests for remote interpreting have:
    - Increased
    - Stayed the same
    - Decreased
    - I do not know
133. Requests for onsite interpreting have:
    - Increased
    - Stayed the same
    - Decreased
    - I do not know
134. The working conditions for remote interpreting have
    - Adapted and improved
    - Not been adapted
    - I do not know

135. What is the source of funding for the interpreting services in the following situations? Tick all that apply per setting

	Government/ agency	Employer	Educational program	Theatre	Deaf Club	Deaf person	Other	N/A
Work	❑	❑	❑	❑	❑	❑	❑	❑
Educational	❑	❑	❑	❑	❑	❑	❑	❑
Medical	❑	❑	❑	❑	❑	❑	❑	❑
Legal	❑	❑	❑	❑	❑	❑	❑	❑
Theatre/performance	❑	❑	❑	❑	❑	❑	❑	❑

136. How much does a freelance interpreter (on average) earn (in euros, excluding taxes)?Per interpreting hour?
137. For traveling time
138. For traveling expenses
139. Do deaf and hearing interpreters receive the same payment?
    - Yes
    - No
    - Other

140. When interpreting for deafblind persons, do interpreters:

	Yes	No	N/A
Receive extra payment for picking up the deafblind person at their home	❑	❑	❑
Next to interpreting also provide guidance during the assignment for the deafblind person	❑	❑	❑
Always work in teams of minimum two interpreters	❑	❑	❑

141. Who determines the interpreting fee for freelance interpreters? (Tick all that apply)
142. The interpreter
    - The government (by regulation or law)
    - The employer
    - The client (requesting party)
    - Other
143. Does the government have a total annual budget limitation for the interpreting services in your country/region?
    - Yes
    - No
144. If the interpreter works for an agency or employer, what is the average annual salary (in euro, excluding taxes) for a full-time position?
145. Does the interpreting fee increase following the years of interpreting experience?
    - Yes
    - No
146. Is there a minimum fee or minimum time per assignment?
    - Yes
    - No

147. Does or can an interpreter receive extra payment for special settings or skills?

	Yes	No	N/A
Legal (court, police, etc.)	❏	❏	❏
Healthcare (general)	❏	❏	❏
Mental health	❏	❏	❏
Conference	❏	❏	❏
Academic	❏	❏	❏
Higher education	❏	❏	❏
Theatre	❏	❏	❏
TV/Media	❏	❏	❏
Deafblind	❏	❏	❏
Another language than the national spoken or signed language, e.g. English	❏	❏	❏
Other	❏	❏	❏

148. Does or can an interpreter receive extra payment for:

	Yes	No	N/A
Remote interpreting	❏	❏	❏
Web streaming	❏	❏	❏
Recording of the interpretation	❏	❏	❏
Copyright of recorded interpretations	❏	❏	❏

149. Does a freelance interpreter receive any additional benefits (social security, pension, etc.)?
   - Yes
   - No
   - Other
150. Does an interpreter earn reasonable payment for the work he or she does?
   - Yes
   - No
   - Other
151. What is the approx. percentage of interpreters in your country/region working full-time or parttime?
152. Can interpreting be the main occupation (so the interpreter does not need a second job to secure an income)?
   - Yes
   - No
   - Not sure
153. How would you classify the development of the working conditions for sign language interpreters in your country/region during the last four years? They have:
   - Remained the same
   - Improved, because:
   - Deteriorated because

Development of the profession

154. How has the Covid-19 pandemic overall impacted the profession in your country/region?
155. How is AI impacting the SLI profession in your country/region?
156. What are currently the main professional opportunities and challenges for sign language interpreters in your country/region?

Appendix 2 – Overview of facts by country[103]

Albania

Population:	32,771,508
Area:	28,748 sq km
Official spoken language:	Albanian
Sign language:	Albanian Sign Language
Sign language recognized:	2014
Deaf signers:	16,000
Interpreters currently working:	12
Interpreter training program:	-
Interpreter organization:	-

Austria

Population:	9,113,574
Area:	83,871 sq km
Official spoken language:	German
Sign language:	Austrian Sign Language
Sign language recognized:	2005
Deaf signers:	10,000
Interpreters currently working:	170
Interpreter training program:	4
Interpreter organization:	Österreichischer Gebärdensprach- DolmetscherInnen- und Übersezterinnen Verband (ÖGSDV) email: info@oegsdv.at website: www.oegsdv.at

Belgium – Flanders & Wallonia

Population (all of Belgium):	11,758,603
Area:	30,528 sq km
Official spoken languages:	Dutch, French, German
Sign language:	Flemish Sign Language, French Belgian Sign Language
Sign language recognized:	Flemish Sign Language (2006), French Sign Language (2003)
Deaf signers:	6,500 (Flanders), 6,000(Wallonia)
Interpreters currently working:	208 (Flanders), 40 (Wallonia)
Interpreter training program:	1 (Flanders), 1 (Wallonia)
Interpreter organization:	*Flanders*: Beroepsvereniging Vlaamse Gebarentaal Tolken (BVGT) email: info@bvgt.be website: www.bvgt.be *Wallonia*: Association des Interprètes en Langue des Signes de Belgique Francophone (ABILS) email: info@abils.net website: www.abils.net

[103] Source used https://en.wikipedia.org/wiki/List_of_European_countries_by_population

Bosnia and Herzegovina

Population:	3,140,096
Area:	51,197 sq km
Official spoken language:	Bosnian, Serbian, Croatian
Sign language:	Bosnian Sign Language
Sign language recognized:	2018
Deaf signers:	5,000
Interpreters currently working:	0
Interpreter training program:	-
Interpreter organization:	-

Bulgaria

Population:	6,714,560
Area:	110,879 sq km
Official spoken language:	Bulgarian
Sign language:	Bulgarian Sign Language
Sign language recognized:	2021
Deaf signers:	5,000
Interpreters currently working:	70
Interpreter training program:	0
Interpreter organization:	Национална Асоциация на преводачите на жестов език в България email: info.naslib@gmail.com website: www.facebook.com/napjeb/

Croatia

Population:	3,848,160
Area:	56,594 sq km
Official spoken language:	Croatian, Bosnian, Serbian
Sign language:	Croatian Sign Language
Sign language recognized:	2015
Deaf signers:	6,500
Interpreters currently working:	100
Interpreter training program:	-
Interpreter organization:	Croatian Association of Sign Language Interpreters for the Deaf (CASLIFD) email: prevoditeljihzj@gmail.com website: www.prevoditeljihzj.eu

Cyprus

Population:	1,370,754
Area:	9,251 sq km
Official spoken language:	Greek, Turkish
Sign language:	Cypriot Sign Language
Sign language recognized:	2006
Deaf signers:	1,000
Interpreters currently working:	14
Interpreter training program:	-
Interpreter organization:	Cyprus Association of Sign Language Interpreters (CyASLI)

email: cyprusasli@gmail.com

Czech Republic

Population:	10,609,240
Area:	78,867 sq km
Official spoken language:	Czech
Sign language:	Czech Sign Language
Sign language recognized:	1998
Deaf signers:	10,000
Interpreters currently working:	137
Interpreter training program:	2
Interpreter organization:	Česká komora tlumočníků znakového jazyka (ČKTZJ) email: info@cktzj.com website: www.cktzj.com

Denmark

Population:	6,002,507
Area:	43,094 sq km
Official spoken language:	Danish, Faroese, Greenlandic (an Inuit dialect), German (small minority)
Sign language:	Danish Sign Language
Sign language recognized:	2014
Deaf signers:	4,000
Interpreters currently working:	350
Interpreter training program:	1
Interpreter organization:	Foreningen af Tegnsprogstolke (FTT) email: ftt@tegnsprogstolk.dk website: www.tegnsprogstolk.dk

England, Wales & Northern Ireland

Population (all of UK):	64,004,432
Area:	243,610 sq km (incl. Scotland)
Official spoken language:	English, Welsh, Irish, Cornish
Sign language:	British Sign Language (BSL)
Sign language recognized:	2003
Deaf signers:	87,000
Interpreters currently working:	1867
Interpreter training program:	8
Interpreter organization:	Association of Sign Language Interpreters (ASLI) email: office@asli.org.uk website: www.asli.org.uk National Union of British Sign Language Interpreters (NUBSLI) email: communications@nubsli.com website: www.nubsli.com

Estonia

Population:	1,334,232
Area:	45,228 sq km
Official spoken language:	Estonian
Sign language:	Estonian Sign Language
Sign language recognized:	2007
Deaf signers:	2,500
Interpreters currently working:	24
Interpreter training program:	-
Interpreter organization:	Eesti Viipekeeletõlkide Kutseühing (EVK) email: evktky@gmail.com website: www.evkty.ee

Finland

Population:	5,623,330
Area:	338,145 sq km
Official spoken language:	Finnish, Swedish
Sign language:	Finnish Sign Language, Finnish-Swedish Sign Language
Sign language recognized:	1995
Deaf signers:	5,500
Interpreters currently working:	800
Interpreter training program:	4
Interpreter organization:	Kieliasiantuntijat ry email: toimisto@kieliasiantuntijat.fi website: www.kieliasiantuntijat.fi

France

Population:	66,650,804
Area:	549,970 sq km
Official spoken language:	French
Sign language:	French Sign Language
Sign language recognized:	2005
Deaf signers:	100,000
Interpreters currently working:	600
Interpreter training program:	6
Interpreter organization:	Association Francaise des Interprètes et Traducteurs en Langue des Signes (AFILS) email: contact@afils.fr website: www.afils.fr

Georgia

Population:	3,806,671
Area:	69,700 sq km
Official spoken language:	Georgian
Sign language:	Georgian Sign Language
Sign language recognized:	2020
Deaf signers:	2,500
Interpreters currently working:	50
Interpreter training program:	1
Interpreter organization:	-

Germany

Population:	84,075,074
Area:	357,022 sq km
Official spoken language:	German
Sign language:	German Sign Language
Sign language recognized:	2002
Deaf signers:	200,000
Interpreters currently working:	1,200
Interpreter training program:	10

Interpreter organization:
Bundesverband der Gebärdensprachdolmetscherinnen Deutschlands e.V. (BGSD)
email: vorstand@bgsd.de
website: www.bgsd.de

Berufsverband der tauben Gebärdensprachdolmetscherinnen E.V. (tgsd)
email: vorstand@tgsd.de
website: www.tgsd.de

Greece

Population:	9,938,844
Area:	131,957 sq km
Official spoken language:	Greek
Sign language:	Greek Sign Language
Sign language recognized:	2000
Deaf signers:	5,000
Interpreters currently working:	70
Interpreter training program:	-
Interpreter organization:	Somatio Diermineon Ellinikis Noimatikis Glosas (SDENG) email: sdeng.gr@gmail.com website: www.sdeng.org.gr

Hungary

Population:	9,632,287
Area:	93,028 sq km
Official spoken language:	Hungarian
Sign language:	Hungarian Sign Language
Sign language recognized:	2009
Deaf signers:	6,303
Interpreters currently working:	100
Interpreter training program:	2
Interpreter organization:	Jelnyelvi Tolmácsok Országos Szövetsége (JOSZ) email: josz.elnokseg@gmail.com

Iceland

Population:	398,266
Area:	103,000 sq km
Official spoken language:	Icelandic
Sign language:	Icelandic Sign Language
Sign language recognized:	2011
Deaf signers:	300
Interpreters currently working:	29
Interpreter training program:	1
Interpreter organization:	HART email: hartfelagtaknmalstulka@gmail.com

Ireland

Population:	5,308,039
Area:	70,273 sq km
Official spoken language:	Irish, English
Sign language:	Irish Sign Language
Sign language recognized:	2017
Deaf signers:	5,000
Interpreters currently working:	123
Interpreter training program:	1
Interpreter organization:	The Council of Sign Language Interpreters (CISLI) email: cisli.ireland@gmail.com website: www.cisli.ie

Italy

Population:	59,146,260
Area:	301,340 sq km
Official spoken language:	Italian
Sign language:	Italian Sign Language (LIS)
Sign language recognized:	2021
Deaf signers:	60,000
Interpreters currently working:	642
Interpreter training program:	6
Interpreter organization:	Associazione Interpreti di Lingua dei Segni Italiana (ANIOS) email: presidenza@anios.it website: www.anios.it Associazione Nazionale Interpreti di Lingua dei Segni Italiana (ANIMU) email: presidenza@animu.it website: www.animu.it

Kosovo

Population:	1,674,125
Area:	10,887 sq km
Official spoken language:	Albanian, Serbian
Sign language:	Kosovar Sign Language
Sign language recognized:	2010
Deaf signers:	3,000
Interpreters currently working:	16
Interpreter training program:	-
Interpreter organization:	Kosovar Association of Sign Language Interpreters (KASLI) email: shksh.pti@gmail.com

Latvia

Population:	1,853,559
Area:	64,589 sq km
Official spoken language:	Latvian
Sign language:	Latvian Sign Language
Sign language recognized:	2000
Deaf signers:	2,000
Interpreters currently working:	32
Interpreter training program:	1
Interpreter organization:	-

Lithuania

Population:	2,830,144
Area:	65,300 sq km
Official spoken language:	Lithuanian
Sign language:	Lithuanian Sign Language
Sign language recognized:	1995
Deaf signers:	8,000
Interpreters currently working:	120
Interpreter training program:	1
Interpreter organization:	Lietuvių gestų kalbos vertėjų asociacija (LGKVA) email: lgkvasociacija@gmail.com website: www.lgkva.lt

Luxembourg

Population:	680,454
Area:	2,586 sq km
Official spoken language:	Luxembourgish, French, German
Sign language:	German Sign Language
Sign language recognized:	2018
Deaf signers:	250
Interpreters currently working:	3
Interpreter training program:	-
Interpreter organization:	-

Malta

Population:	545,405
Area:	316 sq km
Official spoken language:	Maltese, English
Sign language:	Maltese Sign Language
Sign language recognized:	2016
Deaf signers:	500
Interpreters currently working:	11
Interpreter training program:	-
Interpreter organization:	Assocjazzjoni Interpreti tal-Lingwa Sinjali Maltija (AILSM) email: interpreters.lsm@gmail.com website: https://sites.google.com/view/ailsm

Moldova

Population:	2,996,106
Area:	33,843 sq km
Official spoken language:	Romanian
Sign language:	Limbajului Semnelor din Republica Moldova
Sign language recognized:	-
Deaf signers:	3,500
Interpreters currently working:	11
Interpreter training program:	-
Interpreter organization:	-

Montenegro

Population:	632,729
Area:	13,883 sq km
Official spoken language:	Montenegrin
Sign language:	Crnogorski znakovni jezik
Sign language recognized:	-
Deaf signers:	400
Interpreters currently working:	0
Interpreter training program:	-
Interpreter organization:	-

Netherlands

Population:	18,346,819
Area:	41,543 sq km
Official spoken language:	Dutch
Sign language:	Dutch Sign Language
Sign language recognized:	2020
Deaf signers:	30,000
Interpreters currently working:	597
Interpreter training program:	1
Interpreter organization:	Nederlandse Beroepsvereniging Tolken Gebarentaal (NBTG) email: info@nbtg.nl website: www.nbtg.nl

Norway

Population:	5,618,354
Area:	323,802 sq km
Official spoken language:	Norwegian, Sami languages
Sign language:	Norwegian Sign Language
Sign language recognized:	2022
Deaf signers:	5,500
Interpreters currently working:	650
Interpreter training program:	3
Interpreter organization:	Tolkene I Akademikerforbundet (TiA) email: tia@akademikerforbundet.no website: www.akademikerforbundet.no/tia

Poland

Population:	37,332,000
Area:	312,685 sq km
Official spoken language:	Polish
Sign language:	Polish Sign Language
Sign language recognized:	2011
Deaf signers:	9,267
Interpreters currently working:	236
Interpreter training program:	1
Interpreter organization:	Stowarzyszenie Tłumaczy Polskiego Języka Migowego (STPJM) email: biuro@stpjm.org.pl website: www.stpjm.org.pl

Portugal

Population:	10,411,834
Area:	92,090 sq km
Official spoken language:	Portuguese
Sign language:	Portugese Sign Language
Sign language recognized:	1997
Deaf signers:	60,000
Interpreters currently working:	150
Interpreter training program:	3
Interpreter organization:	Associação Nacional e Profissional da Interpretação - Língua Gestual (ANAPI-LG) email: anapilg.dir@gmail.com website: www.anapilg.pt

Romania

Population:	18,908,650
Area:	238,391 sq km
Official spoken language:	Romanian
Sign language:	Romanian Sign Language
Sign language recognized:	2002
Deaf signers:	24,601
Interpreters currently working:	80
Interpreter training program:	1
Interpreter organization:	Asociaţia Naţională a Interpreţilor Autorizaţi în Limbaj Mimico-Gestual (ANIALMG) email: anialmg_rom@yahoo.com website: www.ailg.ro

Russia

Population:	143,997,393
Area:	17,098,242 sq km
Official spoken language:	Russian
Sign language:	Russian Sign Language
Sign language recognized:	2012
Deaf signers:	12,0528

Interpreters currently working:	600
Interpreter training program:	3
Interpreter organization:	Общественная организация "Обьединения переводчиков жестового языка" (RASLI) email: roo-opgy@mail.ru website: www.rasli.ru

Scotland

Population:	5,546,900
Area:	78,782 sq km
Official spoken language:	English
Sign language:	British Sign Language
Sign language recognized:	2015
Deaf signers:	2,600
Interpreters currently working:	150
Interpreter training program:	2
Interpreter organization:	Scottish Collaborative of Sign Language Interpreters (SCOSLI) email: admin@scosli.org website: www.scosli.org

Serbia

Population:	6,689,039
Area:	77,474 sq km
Official spoken language:	Serbian
Sign language:	Serbian Sign Language
Sign language recognized:	2015
Deaf signers:	12,000
Interpreters currently working:	60
Interpreter training program:	-
Interpreter associations:	The Association of Serbian Sign Language Interpreters (ATSZJ) email: centar@atszj.org.rs website: www.atszj.org.rs

Slovakia

Population:	5,474,881
Area:	49,035 sq km
Official spoken language:	Slovak
Sign language:	Slovak Sign Language
Sign language recognized:	1995
Deaf signers:	4,000
Interpreters currently working:	25
Interpreter training program:	1
Interpreter organization:	2 Slovenská komora tlmočníkov posunkového jazyka (SKTPJ) email: komoraspj@gmail.com

Slovenia

Population:	2,117,072
Area:	20,273 sq km
Official spoken language:	Slovene
Sign language:	Slovenian Sign Language
Sign language recognized:	2002
Deaf signers:	1,021
Interpreters currently working:	72
Interpreter training program:	1
Interpreter organization:	Zavod Združenje tolmačev za slovenski znakovni jezik (ZTSZJ) email: info@tolmaci.si website: www.tolmaci.si

Spain

Population:	47,889,958
Area:	505,370 sq km
Official spoken language:	Spanish
Sign language:	Catalan Sign Language, Spanish Sign Language, Valencian Sign Language
Sign language recognized:	2007
Deaf signers:	100,000
Interpreters currently working:	1,000
Interpreter training program:	2
Interpreter organization:	Federación Española de Intérpretes de Lengua de Signos y Guía-Intérpretes (FILSE) email: filse.org@gmail.com website: www.filse.org

Sweden

Population:	10,656,633
Area:	450,295 sq km
Official spoken language:	Swedish
Sign language:	Swedish Sign Language
Sign language recognized:	1981
Deaf signers:	10,000
Interpreters currently working:	*unknown*
Interpreter training program:	3
Interpreter organization:	Sveriges Teckenspråkstolkars Förening (STTF) email: kontakt@sttf.org website: www.sttf.org

Switzerland

Population:	8,967,408
Area:	41,277 sq km
Official spoken language:	French, German, Italian, Romansh
French speaking part	
Sign language:	French Sign Language
Sign language recognized:	2013 (for the canton of Genève)
Deaf signers:	2,750
Interpreters currently working:	29
Interpreter training program:	-
Interpreter organization:	Association Romande de Interprètes en Langue des Signes (ARILS) email: comite.arils@gmail.com website: www.arils.ch
German speaking part	
Sign language:	German Sign Language
Sign language recognized:	-
Deaf signers:	7,500
Interpreters currently working:	80
Interpreter training program:	1
Interpreter organization:	Berufsvereinigung der Gebärdensprachdolmetscher:innen und übersetzer:innen (BGDÜ) email: vorstand@bgdu.ch website: www.bgdu.ch
Italian speaking part	
Sign language:	Italian Sign Language
Sign language recognized:	yes (on canton level)
Deaf signers:	300
Interpreters currently working:	12
Interpreter training program:	1
Interpreter organization:	Interpreti in Lingua dei Segni Svizzera Italiana (ILISSI) email: ilissi@bluemail.ch

Turkey

Population:	87,685,426
Area:	783,562 sq km
Official spoken language:	Turkish
Sign language:	Turkish Sign Language
Sign language recognized:	2005
Deaf signers:	240,000
Interpreters currently working:	200
Interpreter training program:	1
Interpreter organization:	-

Ukraine

Population:	39,000,000 (official), 29,000,000 (UN estimate 2025)
Area:	603,550 sq km
Official spoken language:	Ukrainian
Sign language:	Ukrainian Sign Language
Sign language recognized:	2004
Deaf signers:	30,000
Interpreters currently working:	200
Interpreter training program:	3
Interpreter organization:	-

Appendix 3 - Overview of all tables and figures

Part 1

Tables

1.1 Year of establishment of independent interpreter associations
1.2 Number of associations and their volunteer and paid positions
1.3 Main issues national associations of sign language interpreters report to face

Figures

1.1 Sign language interpreters and national associations: relative membership (%)
1.2a Total membership numbers by national or regional association (>100 members)
1.2b Total membership numbers by national or regional association (<100 members)
1.3 Percentage of gender distribution among the members of all interpreter associations (%)
1.4 Does your association collect member information on gender, race and ethnicity by asking members to self-identify?
1.5 Representation of ethical and racial minorities in the interpreter associations
1.6 Number of deaf and hard of hearing interpreters as member of an interpreter association
1.7 Individual membership requirements
1.8 Individual membership benefits - number of associations per category
1.9 Interpreter associations' aims - number of associations per category
1.10 Funding of the interpreter association
1.11 National associations - 2025 individual interpreter membership fee (in euros)
1.12 Annual income per association based on individual membership fees (in euros)
1.13 Other membership categories

Part 2

Tables

2.1 Yes, definition of the sign language interpreter profession in the national legislation
2.2 Countries/regions with an official SLI accreditation/registration body
2.3 Countries/regions without an official SLI accreditation/registration body
2.4 Overview of prerequisites and criteria per country to register as a sign language interpreter
2.5 Countries with a complaints' procedure - parties that can file a complaint
2.6 Alternative mechanisms through which interpreters receive client feedback
2.7 Consequences following a valid complaint

Figures

2.1 Does your association focus on quality assurance?
2.2 Is the sign language interpreter profession regulated in your country?
2.3 Are there other professionals in your country/region who are put in place instead of qualified interpreters?
2.4 Registered interpreters in countries & regions with a registry
2.5 Number of interpreters accredited or registered to work with deafblind persons
2.6 Are deaf clients aware of the complaints' procedure?
2.7 Are hearing clients aware of the complaints' procedure?
2.8 When deaf clients are not happy with the services received, do they use the possibility to file a complaint?
2.9 When hearing clients are not happy with the services received, do they use the possibility to file a complaint?

Part 3

Tables

3.1 Opportunities and barriers for deaf interpreters
3.2a How hearing interpreters support the work of deaf interpreters
3.2b How do hearing interpreters support the work of deaf interpreters
3.3 Are deaf interpreters included in official standards, qualification frameworks, or professional guidelines in the interpreting sector?
3.4 Inclusion and exclusion of deaf interpreters in official frameworks
3.5 Overall measures needed to strengthen the position of deaf interpreters
3.6 Number of deaf interpreters working as interpreters
3.7 Options deaf interpreters have when co-interpreting
3.8 Main barriers deaf interpreters face in recruitment as an interpreter
3.9 Overall comments on the work of deaf interpreters
3.10 Overview of available training deaf interpreters can enroll in
3.11 Countries reporting no or very limited training for deaf interpreters
3.12 Number of trained deaf interpreters
3.13 Rating by deaf interpreters

Figures

3.1 Term mostly used in the countries/region when referring to sign language interpreters who are deaf
3.2 How has the profession of sign language interpreters developed for deaf interpreters in your country/region during the last five years?
3.3 Number of deaf interpreters in Europe
3.4 Settings deaf interpreters are most recruited for
3.5 How deaf interpreters work
3.6 Organization of the deaf interpreters' payment
3.7 Distribution of assignments by deaf interpreters
3.8 How deaf interpreters are trained
3.9 Are deaf interpreters trained to interpret via live captions and/or as co-interpreter with a hearing interpreter?

Part 4

Tables

4.1 Number of interpreter training possibilities
4.2 Countries that do not have formal interpreter training programs
4.3 Overview of sign language interpreter training programs in Europe
4.4 Sign language interpreter training programs and deaf students
4.5 Prerequisites to attend educational programs
4.6 Training on interpreting for deafblind persons
4.7 An interpreting degree as a prerequisite to work as an interpreter
4.8 Requirements of CPD / further training
4.9 Specializations available and/or formally recognized
4.10 Availability of mentorship program for (student/trainee) sign language interpreters
4.11 Available mentorship program or supervision for (student/trainee) sign language interpreters

Figures

4.1 Number of educational programs in Europe per educational level
4.2 Prerequisites to attend the educational program for SLI

4.3 Number of training program that educated students in specific topics
4.4 Number of programs that teach additional languages
4.5 Educational programs and if they have an internship or traineeship
4.6 Is continuing education mandatory to work as an interpreter?
4.7 Number of countries/regions with specializations

Part 5

Tables

5.1 EU member states 2025, 2020, 2016 - working sign language interpreters
5.2 Total number of sign language interpreters in fifteen countries
5.3 Right to a sign language interpreter
5.4 From what age onward does a deaf or deafblind person have the right to an interpreter in an educational setting?
5.5 Possibility for a deaf, hard of hearing or deafblind person to request an interpreter in a specific setting
5.6 WORK SETTING: Party responsible for the payment of the interpreting services
5.7 EDUCATIONAL SETTING: Party responsible for the payment of the interpreting services
5.8 MEDICAL SETTING: Party responsible for the payment of the interpreting services
5.9 LEGAL SETTING: Party responsible for the payment of the interpreting services
5.10 THEATRE SETTING: Party responsible for the payment of the interpreting services
5.11 Recognition of sign language per country or region
5.12 Employment forms per country and region
5.13 Overview per country of payment for freelance interpreting
5.14 Extra pay for specialties and specific settings
5.15 Interpreting for deafblind persons - the interpreter
5.16 Who determines the fee for freelance interpreters?
5.17 Does the government have a total annual budget limitation for the interpreting services in your country/region?
5.18 Classification of the development of the working conditions for sign language interpreters during the last five years (2020-2025)
5.19 Impact of Covid19 pandemic on the sign language interpreter profession
5.20 AI impact on sign language interpreter profession
5.21 Main professional challenges and opportunities in 2025

Figures

5.1 Europe 2025 - working sign language interpreters
5.2 EU member states 2020 > 2025 working sign language interpreters
5.3 Interpreters working full-time vs. part-time (%)
5.4 Number of sign language interpreters interpreting for deafblind persons
5.5 Sufficient number of interpreters to meet the demand?
5.6 If there is a lack of interpreters, where is this lack noticeable
5.7 Deaf signers (45 countries & regions)
5.8 Deaf signers in the EU
5.9 Deaf signers per interpreter (43 countries & regions)
5.10 Who pushes mostly for action to ensure the use of the legal right to an interpreter?
5.11 Options to request a sign language interpreter (47 countries & regions)
5.12 Countries - forms of employment (%)
5.13 How interpreters mostly work in the countries
5.14 Since the Covid19 pandemic changes in the requests for remote interpreting
5.15 Change in working conditions for remote interpreting since the Covid19 pandemic

5.16 Freelance interpreters - Range of minimum fees in euros per hour per country/region
5.17 Minimum interpreting fee per freelance hour in 2025
5.18 Extra pay for specific settings / competencies in number of countries
5.19 Extra pay for additional requirements in number of countries
5.20 Interpreter interpreting for a deafblind person
5.21 Does an interpreter earn reasonable payment for the work they do?
5.22 Reasonable pay for interpreters
5.23 Can interpreting be the main occupation? Year 2020 vs 2025
5.24 Can interpreting be the main occupation, or is a second job needed to secure income
5.25 Does the government have annual budget limitations on interpreting fees?
5.26 Interpreter working conditions per country/region during the last five years
5.27 Impact of Covid19 pandemic on the profession
5.28 Impact of AI on the profession
5.29 Challenges and opportunities in the profession in the countries in 2025

This book has been produced in line with the EU GPSR guidelines about the safety of products.

The General Product Safety Regulation is the European Union's updated framework for ensuring that all consumer products, including books, are safe for consumers.

This book has been printed by Libri Plureos GmbH. The printer has issued safety certificates for the materials - like ink, paper and glue - being used.

The product identifier is: 9789403895536

The author is responsible for the content of the book and had it produced by Bookmundo.

Should there be any questions in regard to the safety of the product, please contact us.

Bookmundo
Delftsestraat 33
3013AE Rotterdam
The Netherlands
info@bookmundo.com